Foundations of Spectroscopy

Simon Duckett

Department of Chemistry, University of York

Bruce Gilbert

Department of Chemistry, University of York

Series sponsor: **ZENECA**

ZENECA is a major international company active in four main areas of business: Pharmaceuticals, Agrochemicals and Seeds, Specialty Chemicals, and Biological Products.

ZENECA's skill and innovative ideas in organic chemistry and bioscience create products and services which improve the world's health, nutrition, environment, and quality of life.

ZENECA is committed to the support of education in chemistry and chemical engineering.

OXFORD
UNIVERSITY PRESS

D0073810

OXFORD

UNIVERSITY PRESS

Great Clarendon Street, Oxford OX2 6DP

Oxford University Press is a department of the University of Oxford.
It furthers the University's objective of excellence in research, scholarship,
and education by publishing worldwide in

Oxford New York

Athens Auckland Bangkok Bogotá Buenos Aires Calcutta
Cape Town Chennai Dar es Salaam Delhi Florence Hong Kong Istanbul
Karachi Kuala Lumpur Madrid Melbourne Mexico City Mumbai
Nairobi Paris São Paulo Singapore Taipei Tokyo Toronto Warsaw

with associated companies in Berlin Ibadan

Oxford is a registered trade mark of Oxford University Press
in the UK and in certain other countries

Published in the United States
by Oxford University Press Inc., New York

A catalogue record for this book is available from the British Library

Library of Congress Cataloging in Publication Data

ISBN   0 19 850335 0 (Data applied for)

Typeset by EXPO Holdings, Malaysia

Printed in Great Britain
on acid-free paper by Bath Press, Bath, Avon

Series Editor's Foreword

Oxford Chemistry Primers are designed to provide clear and concise introductions to a wide range of topics that may be encountered by chemistry students as they progress from the freshman stage through to graduation. The Physical Chemistry series contains books easily recognized as relating to established fundamental core material that all chemists need to know, as well as books reflecting new directions and research trends in the subject, thereby anticipating (and perhaps encouraging) the evolution of modern undergraduate courses.

In this Physical Chemistry Primer, Simon Duckett and Bruce Gilbert present an exceptionally clearly written and elegant introductory account of the *Foundations of Spectroscopy*. The book explains in simple terms the basic ideas and applications of a subject which is essential knowledge for all practising chemists. This Primer will be of interest to all students of chemistry and their mentors.

Richard G. Compton
Physical and Theoretical Chemistry Laboratory,
University of Oxford

Preface

This book is written primarily for those studying first-year University courses in Chemistry and for those preparing to do so.

It is designed to reflect significant recent advances in the use of spectroscopic and diffraction methods, not only for obtaining an analysis of elements and groups present in a molecule but also for establishing the arrangement of the constituent atoms. These developments have had a profound effect by increasing scientific knowledge in the fields of chemistry and molecular biology, reflected in the elucidation of the structure and function of a wide range of compounds including drugs, proteins and enzymes, and nucleic acids.

It is important that such work and its appreciation should find its place in the curriculum — as a means of educating chemistry students about essential principles and wide-ranging applications and to show how problem-solving skills are developed and applied in industrial and research environments. We also hope to convey the enjoyment and satisfaction of successful spectrum analysis.

We have included mass spectrometry and X-ray diffraction, along with traditional spectroscopic techniques; the former is the method *par excellence* for molecular mass and formulae determination and the latter provides details of molecular structure, providing information complementary to i.r., n.m.r. and u.v.-visible spectroscopy. We introduce the essential physical principles of each method, many examples of spectral analysis, and some problems; further reading and practice are strongly encouraged.

SI units have been adopted, with IUPAC nomenclature; trivial names accompany the recommended names in parentheses. Accurate mass data are taken from *Mass and Abundance Tables for Use in Mass Spectrometry* by J. H. Beynon and A. E. Williams, Elsevier, Amsterdam, 1963, and fragmentation patterns from *Compilation of Mass Spectral Data* by A. Cornu and R. Massot, Heyden, London, 1966.

We thank especially the following for assistance in recording spectra: Kin Mya Mya, Anthony Crawshaw, Zygmunt Derewenda, Guy Dodson, Chris Hall, Reuben Girling, Rod Hubbard, Robert Liddington, and Ted Parton. We acknowledge permission to use Fig. 6.19 (The Royal Society) and Fig. 5.29 (one of our students). We are grateful for particularly relevant advice from schoolteachers including David Bevan, Michael Cane, Peter Gradwell, Geoff Liptrot, Bill Pickering, and George Walker. Finally, our special thanks go to David Waddington and Barry Thomas for enthusiasm and encouragement, and to Sue Street and Adrian Whitwood for assistance in producing the manuscript.

York   S. B. D. and B. C. G.
1999

Contents

Mass spectrometry

he technique of mass spectrometry, which owes its origin to pioneering
xperiments carried out at the beginning of the twentieth century, is now
.tablished as a means for obtaining the formulae and structures of mole-
des. It enables many sophisticated structural problems to be solved rapidly,
ven when only minute quantities of material are available.

.1   The mass spectrometry experiment

he principle of the method is to obtain a positively charged ion characteris-
 of the substance under investigation, and then to determine the *mass* of
is ion using an approach closely related to that employed by J. J. Thomson
r measuring the charge-to-mass ratio (z/m) for electrons. The procedure
volves the use of electric and magnetic fields to deflect the charged
rticles.

   Thomson used a magnetic field to deflect a beam of positive ions, obtained
 the ionization (loss of an electron) of neon atoms. Close examination of
e trace produced by the positive ions as they impinged on a detector
monstrated that there were two different types of ion, characterized as
ose from the two neon **isotopes** (^{20}Ne and ^{22}Ne), which differ in mass
cause of the different numbers of neutrons in their nuclei. The design of a
mple mass spectrometer is shown in Fig. 1.1. A very small amount of the
pour of the substance to be studied (obtained, for example by heating the
mple) is introduced into the ionization chamber at very low pressure (about
)$^{-4}$ N m^{-2}). The vapour is bombarded with high-energy electrons, and the
ollision between an electron and the molecule M (or atom) causes an elec-
on to be ejected, leaving a positively charged ion (M^+). The ions are
tracted by an applied electrostatic potential and are hence accelerated
wards the negative plate. The ions pass through a slit in the plate and into a
agnetic field: the positive ions then become deflected by an amount which
pends upon their *mass* (m) and the *charge* (z). The lighter the ion and the
eater its charge, the greater will be the deflection.

Ionisation

$M \rightarrow M^+ + e^-$

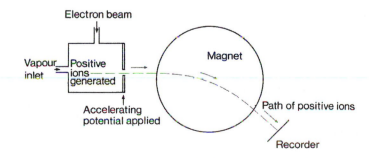

Fig. 1.1 Basic features of a simple mass spectrometer

The derivation of the exact relationship is as follows: for an acceleration potential V the *potential energy* of an ion, of charge z, generated in the ion chamber is zV. The ion is accelerated through the slit, and in this process its potential energy zV is completely converted into *kinetic energy* $\frac{1}{2}mv^2$, where m and v are the mass and velocity, respectively, of the ion (eqn 1.1). When the positive ion passes into the magnetic field (of magnetic flux density B) it experiences a force at right angles both to the direction of motion and to the field direction. The magnitude of this force is Bzv. The positive ion is now constrained to move in the arc of a circle of radius R, as given in equation 1.2.

$$zV = \frac{mv^2}{2} \qquad (1.1)$$

$$Bzv = \frac{mv^2}{R} \qquad (1.2)$$

Combination of these equations (eqn 1.3) leads to the important expression linking m, z, B, R and V. This equation shows that for an ion of given mass (m) and charge (z) the radius of the circle of motion (R) is determined by B and V, i.e. the magnitudes of the magnetic and electric fields. In practice, R remains fixed by the geometry of the apparatus and the position of the detector. It can then be seen that if V is kept constant and B is varied, equation 1.3 will be satisfied for ions of different m/z for different values of B. The value of B needed to get a particular type of ion to be deflected to the recorder is a measure of m/z for that ion.

$$\frac{m}{z} = \frac{B^2R^2}{2V} \qquad (1.3)$$

Most positive ions generated will have lost just one electron and they will therefore have the same charge (opposite in sign, but equal in magnitude, to that of the electron). This means that, as B is varied, ions of different mass arrive at the recorder and a spectrum of the masses of the various ions concerned can be plotted. Since the mass of the electron is very small compared with the mass of the nucleus, the experiment is effectively determining the masses of the parent molecules.

Early mass spectrometers like this achieved a low resolution (e.g. separation of two fairly similar masses, such as ^{20}Ne and ^{22}Ne). Many modern mass spectrometers are designed with an extra focusing system which enables increased resolution to be obtained. As before, a narrow beam of positive ions with a small but finite range of kinetic energies is produced; this spread of energies must be reduced for more precise work so that equation 1.3 is strictly applicable. This is achieved by passing the ions through an electric field (an electrostatic analyser) which deflects the ions according to their kinetic energies (Fig. 1.2). Then only one small component of the resulting beam, with a well-defined kinetic energy, is passed into the magnetic field for focusing of ions of given m/z values. As before, a scan of mass is obtained by varying B (although in certain circumstances it is possible to obtain a more rapid scan by varying V, keeping B constant). An electron-multiplier usually serves as a collector and detector of positive ions and the arrival of the ions gives rise to a signal. The signal is subsequently displayed as a peak on a oscilloscope or recorder or is processed (and stored) by an on-line computer. This procedure produces a **mass spectrum** — effectively a plot of the masses of the positive particles present against the relative number of ions of each mass. The scan is calibrated with a peak from a substance of known relative molar mass. The resulting spectrum can usually be obtained from less than a nanogram of material in a few seconds.

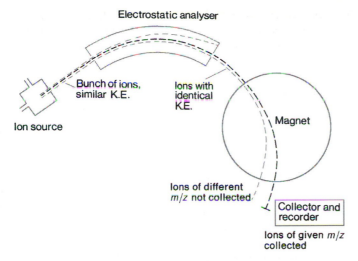

Fig. 1.2 Diagram of a double-focusing mass spectrometer

.2   Measuring relative molecular and atomic masses

Modern mass spectrometers with a double-focusing facility can be used in various kinds of study. For example, they may be employed to give fairly rapid scans of the relative masses of the ions from a variety of substances. Alternatively, under conditions of high resolution, they can be used to separate closely spaced peaks and to determine the appropriate relative atomic and molar masses with precision. The following example illustrates the advantages of the latter approach if the maximum amount of information is to be derived. A peak which corresponds to mass 28 might be due to nitrogen ($^{14}N_2$), carbon monoxide ($^{12}C^{16}O$), or ethene ($^{12}C_2{}^1H_4$). However, these three molecules have slightly different relative molecular masses, as shown (these are based on the internationally accepted scale, with 12 exactly for the ^{12}C isotope).

A high resolution mass spectrometer can readily be used to identify a peak exactly enough for it to be characterized as the positive ion from one of these. Further, if all three substances were to be present together, then under high-resolution conditions three separate peaks could be resolved. If one relative molar mass is accurately known, this can be used to calibrate the field scan so that the other molecular masses can also be accurately determined. Relative weights of separate peaks can also be used to obtain quantitative information. For example, from the mass spectrum of neon can be measured not only the relative atomic masses of the constituent isotopes (^{20}Ne, ^{21}Ne, ^{22}Ne), to an accuracy of 1 part in 10^6, but also the relative abundance of the separate isotopes in the mixture.

Remember to distinguish the *separate* isotopic atomic masses measured with the mass spectrometer from the *weighted average* obtained by other (chemical) methods. For example, ^{35}Cl has a relative atomic mass of 34.9688, and that of ^{37}Cl is 36.9659; the average atomic mass of the natural mixture of isotopes (75.53% ^{35}Cl, 24.47% ^{37}Cl) is 35.45.

Relative Molar Masses

$^{14}N_2$	28.0062
$^{12}C^{16}O$	27.9949
$^{12}C_2{}^1H_4$	28.0313

Relative atomic mass (%)	Relative abundance
^{20}Ne 19.9924	90.92
^{21}Ne 20.9940	0.26
^{22}Ne 21.9914	8.82

1.3   Mass spectrometry of molecules

When an organic compound is introduced into the mass spectrometer, the molecules become ionized, by the loss of an electron, and the positive ion produced pass through the focusing system, leading usually to a peak at the appropriate relative molar mass. However, the mass spectrum of an organic compound also contains extra information about fragmentation which can be extremely useful.

Fig. 1.3 shows the mass spectrum of ethanol (CH_3CH_2OH): this is a print out of signal height (proportional to the number of ions of given m/z) against increasing m/z (since nearly all the ions have the same unit charge this axis effectively corresponds to increasing mass).

Peaks occur at (or close to) most integral values (the extra precision possible with high resolution is not usually employed at this stage). Many of the peaks are derived from ethanol by processes which will be described shortly.

There may also be peaks due to traces of air in the instrument: this gives rise to signals from N_2 (m/z 28) and O_2 (m/z 32), approximately in the expected ratio 4:1. These peaks may be used to calibrate the scan. The peak heights are expressed as percentages of the height of the highest peak (called the **base peak**), which in this example is the peak with m/z 31.

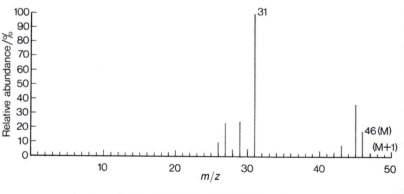

Fig. 1.3 Mass spectrum of ethanol

The spectrum shows the expected peak at m/z 46, corresponding to the **molecular ion** (M^+) of the parent molecule (the relative molecular mass is the sum of the relative atomic masses of $2C + 6H + O$).

There is also a very small peak at m/z 47, called the (M + 1) peak, which corresponds to the relatively few ethanol molecules present which, because they contain a ^{13}C, ^{17}O, or ^2H atom, have a molar mass of 47. ^{13}C has a natural abundance compared with ^{12}C of 1.1%; for ^{17}O, relative to ^{16}O, the figure is 0.04% and for ^2H, relative to ^1H, the abundance is 0.01%. An even smaller (M + 2) peak arises from the molecules which contain two ^{13}C atoms or an ^{18}O atom, or a ^{13}C atom and an ^{17}O atom etc.

Other peaks in the spectrum occur because some ethanol molecules which are ionized to give M^+ then *fragment* to give smaller positive ions, a process which is understandable in terms of the high energy of the bombarding elecrons. The positive fragments are also accelerated and focused to be collected and recorded for their particular values of m/z. The large peak in the mass spectrum of ethanol (the base peak) is at m/z 31: this corresponds to the fragment $[CH_2OH]^+$ obtained by loss of CH_3 from the parent ion $[CH_3CH_2OH]^+$. The structure of positive ions will be discussed later, as will some guidelines for interpreting fragmentation patterns, but it should at this stage be apparent that these peaks contain important clues about the molecular structure.

1.4 Analysis of mass spectra

The molecular ion

Many molecules give a peak of appreciable size for the molecular ion. It is usually a reliable guide that a molecule with π-electrons (e.g. benzene) will give a detectable molecular ion (M) since one of these electrons can normally be lost (to give M^+) without the breakdown of the bonding framework in the molecule. However, because in some cases no peak from a molecular ion can be observed, care must be taken before it can be assumed that the peak at highest m/z is from the molecular ion.

It may be possible at this stage to determine very accurately the *relative molecular mass* of any given peak (if the high-resolution facility is available) and this will then be carried out for the molecular ion itself. Because different atoms do not have exactly integral atomic masses and, in addition, because various combinations of similar mass are not identical (contrast for example C_2H_4 and N_2) the exact relative molecular mass (to 3 or 4 places of decimals) characterizes the molecular formula exactly. For instance, a molecular peak with m/z at almost exactly 60 could be from various possible molecules with different formulae, including $C_2H_4O_2$ [e.g. ethanoic acid (acetic acid, CH_3CO_2H)] and C_3H_8O (e.g. propanol). Under conditions of high resolution, these possibilities can be clearly distinguished. Indeed, given a precisely determined relative molecular mass we can obtain the molecular formula: for example, a molecular ion with m/z 94.0419 can be reliably attributed to a compound with the molecular formula C_6H_6O.

Formula	Relative molar mass
$C_2H_4O_2$	60.0211
C_3H_8O	60.0575

Information can often be extracted from the M, (M + 1), and (M + 2) peaks even if a high resolution facility is not available. For example, for ethanoic acid ($C_2H_4O_2$), the height of the (M + 1) peak at m/z 61, which is mainly due to $^{12}C^{13}CH_4O_2$, should be just over 2% of the height of the peak from the molecular ion (m/z 60). This is because there is approximately a 2.2% chance that a molecule will contain one ^{13}C atom (there will be a much smaller percentage of molecules containing ^2H or ^{17}O). The corresponding figure for C_3H_8O is just over 3%, and for a compound with, say, eleven carbon atoms the relative intensities of M: (M + 1) peaks should be about 88:12 (i.e. [100 − (11 × 1.1)]:[11 × 1.1]). Clearly, then, measurement of the *relative heights* of the M and (M + 1) peaks, and sometimes of the (M + 2) peak can be diagnostically useful, and extensive tables of accurate M:(M + 1):(M + 2) ratios for various molecular formulae are available. In any case, from a brief inspection sensible deductions can usually be made. Thus the (M + 1) peak will be approximately N % of the main peak if the formula has N carbon atoms, a larger-than-usual (M + 2) peak may indicate that a sulfur atom is present in the molecule (^{34}S has a natural abundance of 4.22%), and so on.

In certain molecules the effect can be particularly striking. Fig. 1.4 shows the mass spectrum of chloromethane in which the peaks at m/z 50 and 52 are from the molecular ions of $CH_3{}^{35}Cl$ and $CH_3{}^{37}Cl$, respectively, their relative intensities being in the ratio expected from the relative isotopic abundances of ^{35}Cl and ^{37}Cl (approximately 3:1). For bromomethane (Fig. 1.5) the two almost equally intense peaks are from $CH_3{}^{79}Br$ and $CH_3{}^{81}Br$, the two bromine isotopes having approximately the same abundance.

Another useful rule is that a molecular ion with an *odd* value of m/z generally characterizes a molecule with an odd number of nitrogen atoms.

It should also be remembered that a peak of reasonable intensity at the highest m/z value observed is not necessarily from the molecular ion, but may be instead from part of the fragmentation pattern of a compound whose molecular ion has a peak too small to be clearly established. Wherever possible, therefore, data should be interpreted together with information from other spectroscopic techniques and from conventional molecular mass and both empirical and molecular formulae determination.

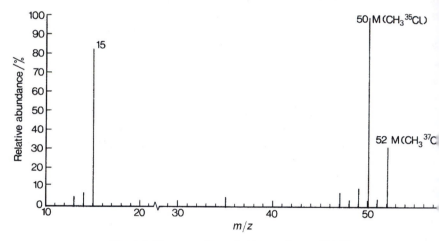

Fig. 1.4 Mass spectrum of chloromethane, CH_3Cl

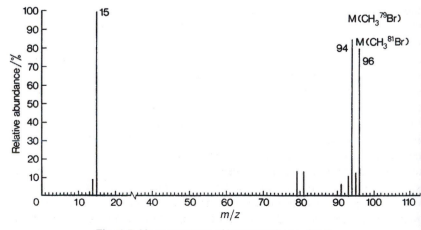
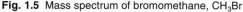

Fig. 1.5 Mass spectrum of bromomethane, CH_3Br

Fragmentation patterns

A variety of fragmentation pathways is normally possible for M^+ and for each route one of the fragments retains the positive charge. For example, another possibility here is fragmentation to $P + Q^+$, and further fragmentation of either P^+ or Q^+ may also occur.

The recognition of preferred modes of fragmentation (e.g. whether for this molecule a larger peak for P^+ or Q^+ is obtained) is assisted by practice with spectra from known molecules, but deductions are based mainly on chemical intuition, and a few simple guidelines. Most of the principles involved in recognizing and predicting fragmentation patterns are closely related to those employed for discussing the chemistry of reactions in solution. For instance, we will need to consider which of a variety of possible fragments is best able to bear a positive charge, which bond is the weakest and therefore most likely to break, and which stable entities might readily be formed in simple decomposition pathways.

First, remember to do some elementary book-keeping with electrons and charges; almost all molecules have an even number of (paired) electrons, so that the positive molecular ion must have not only a *charge* but also an odd number of electrons. The fate of both the charge and the unpaired electron should be considered when fragmentation patterns are being interpreted.

The main types of fragmentation possible for a molecule may be characterized as follows.

Fragmentation

$$M \rightarrow M^+ \rightarrow P^+ + Q$$

(i) *Simple cleavage.* This involves the breakage of a single bond in the molecular ion, and a good example is provided by the mass spectrum of ethanol (Fig. 1.3). The molecular ion is at m/z 46 and the base peak is at m/z 31; the latter corresponds to a molecular ion which has lost a group of mass 15 before being accelerated and focused. It is described as an (M-15) peak, and is due to the ion $[CH_2OH]^+$ formed as shown.

Note that the 'dot' indicates an unpaired electron. The products of this fragmentation then are the charged *ion* $[CH_2OH]^+$ (since the charge resides effectively on carbon this type of species is sometimes called a carbonium ion or carbocation and written $^+CH_2OH$) and the neutral methyl radical $^·CH_3$. Only the former, being charged, is recorded in the mass spectrum. Some of the other peaks arise as shown. The peaks with m/z 27–29 are typical of a molecule containing an ethyl group, just as the appearance of an (M-15) peak is usually indicative of a methyl group in the parent compound.

The reason that the peak with m/z 31, $[CH_2OH]^+$, is larger than those of the other positive ions is that this is a relatively *stable* positive ion compared to some of the other possible ions (e.g. $[CH_3]^+$, $[CH_3CH_2]^+$). This stability arises because oxygen has a lone-pair of electrons which can help to stabilize the positive charge on the adjacent carbon atom: this is possible because there is a spreading (delocalization) of both the charge and the electrons between carbon and oxygen, a phenomenon which is indicated diagrammatically as shown. The use of the double-headed arrow implies that the actual electronic structure is somewhere in between the two extremes indicated.

A similar explanation accounts for the formation of the fairly intense peak with m/z 45, attributed to the ion $^+CH(CH_3)OH$ formed by loss of a hydrogen atom: this ion is more stable than $CH_3CH_2O^+$. This mode of fragmentation is

$$[CH_3CH_2OH]^{+\cdot} \quad m/z\ 46$$

$$\downarrow$$

$$^·CH_3 + [CH_2OH]^+$$
$$m/z\ 31$$

$$[C_2H_5OH]^{+\cdot} \longrightarrow H^· + [C_2H_5O]^+$$
$$m/z\ 45$$

$$[C_2H_5OH]^{+\cdot} \longrightarrow HO^· + [C_2H_5]^+$$
$$m/z\ 29$$

$$[C_2H_5OH]^{+\cdot} \longrightarrow H_2O + [C_2H_4]^{+\cdot}$$
$$m/z\ 28$$

$$[C_2H_5]^+ \longrightarrow H_2 + [C_2H_3]^+$$
$$m/z\ 27$$

$$^+CH_2-\overset{..}{\text{O}}-H \longleftrightarrow CH_2=\overset{..}{\text{O}}-H$$

often observed when an ion can be produced with the positive charge on a carbon atom adjacent to an atom with a lone pair of electrons (e.g. O, S, N), and is often significant for alkanols, ethers, thiols, and amines. Their typical fragmentation patterns often provide a means by which these molecules can be recognized.

Other cases where stabilized positive ions are produced include compounds containing the phenylmethyl (benzyl) group and also carbonyl-containing compounds. For example, Fig. 1.6 is the mass spectrum from methylbenzene (toluene) which illustrates the behaviour of compounds in the former group. In addition to the molecular ion at m/z 92 and the (M + 1) peak at m/z 93 (which has an intensity which is about 8% of the molecular peak, since there are seven carbon atoms in methylbenzene; see page 5), there is an intense peak (the base peak) at m/z 91. This is from some of the methylbenzene molecules in the ionization chamber which lose first an electron and then a hydrogen atom, to give the phenylmethyl cation as shown.

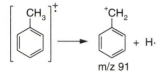

m/z 91

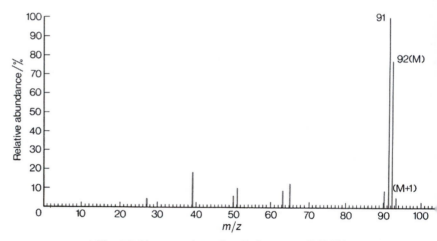

Fig. 1.6 Mass spectrum of methylbenzene, $C_6H_5CH_3$

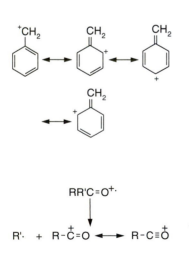

The reason for the comparative stability of this cation is the ease with which the *aromatic ring* can delocalize the positive charge, a phenomenon which can be represented diagrammatically as shown.

This type of fragmentation, to give a peak with $m/z = 91$, is characteristic of compounds of the type $C_6H_5CH_2X$. The mass spectrum from methylbenzene also shows other small peaks which indicate that alternative modes of fragmentation after high-energy bombardment are possible. These include the formation of two-, three-, and four-carbon fragments (that with m/z 51 is $[C_4H_3]^+$) from the aromatic ring; the occurrence of these, though helpful, is not as diagnostically useful as the evidence from the main pathway.

Carbonyl-containing compounds RC(O)R′ tend to decompose to give fragment ions in which the positive charge is again shared between carbon and oxygen (RCO⁺). Other peaks arise because fragmentation occurs at the other C-alkyl bond to give R′CO⁺ and because these ions readily lose carbon monoxide. For example, the mass spectrum from butanone (methyl ethyl ketone, Fig. 1.7) shows the molecular ion (m/z 72) and the peaks at m/z 57

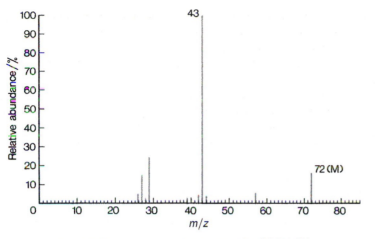

Fig. 1.7 Mass spectrum of butanone, $CH_3COCH_2CH_3$

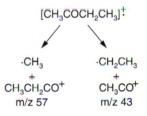

(M-15) and 43 (M-29), diagnostic of loss of $\cdot CH_3$ and $\cdot CH_2CH_3$ from the parent positive ion. The peak at m/z 29 ($^+CH_2CH_3$) evidently arises via loss of CO from $CH_3CH_2CO^+$. The peaks at 57 (M-15) and 29 (M-43) help characterize a methyl-substituted ketone (loss of $\cdot CH_3$ and both $\cdot CH_3$ and CO, respectively).

The spectrum of dimethylbutanone (methyl t-butyl ketone), Fig. 1.8, shows the (M-43) peak quite clearly; the dimethylethyl (t-butyl) carbonium ion $^+C(CH_3)_3$, m/z 57, gives a particularly intense peak because a *tertiary* carbonium ion (with three alkyl groups attached to the carbon bearing the positive charge) is formed. This type of fragmentation is favoured here because a tertiary carbonium ion is more stable than a *secondary* carbonium ion, which is more stable than a *primary* carbonium ion (because of the overall electron-releasing property of alkyl groups). This means that fragmentation is generally preferred at the point of branching.

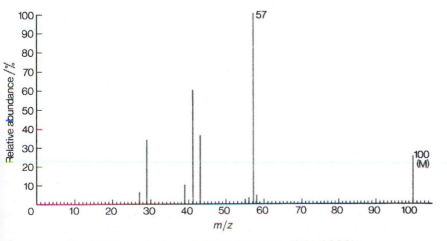

Fig. 1.8 Mass spectrum of dimethylbutanone, $(CH_3)_3CCOCH_3$

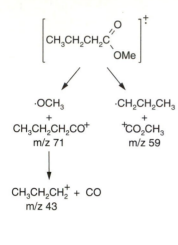

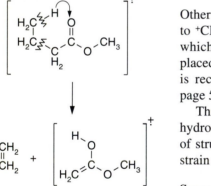

(ii) *Fragmentation with rearrangement.* Occasionally, a fragmentation process is detected which is rather more complicated than those discussed in section (i) because molecular rearrangements are involved.

An example is provided by the mass spectrum of methyl butanoate ($CH_3CH_2CH_2CO_2CH_3$), which is shown in Fig. 1.9. There is a trace of the expected molecular peak at m/z 102, and cleavage of the bonds to the carbonyl group leads to the peaks at m/z 71 (loss of $\cdot OCH_3$), and subsequently 43 (via loss of CO), and 59 (loss of $\cdot CH_2CH_2CH_3$).

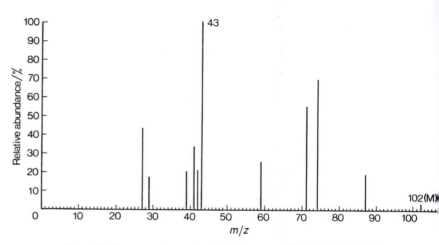

Fig. 1.9 Mass spectrum of methyl butanoate, $CH_3CH_2CH_2CO_2CH_3$

Other fragmentations lead to loss of methyl to give the ion with m/z 87 and to $^+CH_2CH_3$ (m/z 29). However, the unusual peak is that at m/z 74 (M-28) which is thought to arise by transfer of a hydrogen atom to the suitably placed oxygen at the same time as fragmentation [the product ion (m/z 74) is recognizable as the enol form of methyl ethanoate, $CH_3CO_2CH_3$: see page 51].

This is known as a **McLafferty** rearrangement and tends to occur when a hydrogen atom and a carbonyl oxygen come into close proximity. Inspection of structural models reveals that this can be achieved with the minimum of strain when there are six atoms in the chain.

Some examples of fragmentation patterns

The simple rules laid down so far should provide assistance with the solving of a variety of mass spectra. The following examples to a considerable extent typify the class of organic compound to which they belong.

Chloroethane, (Fig. 1.10). The mass spectrum shows the two expected molecular ions from $C_2H_5{}^{35}Cl$ and $C_2H_5{}^{37}Cl$, at m/z 64 and 66, respectively. The other main peaks are formed as shown. The loss of $\cdot H$ and $\cdot CH_3$ in the first two paths leaves the positive charge next to the chlorine [cf. $^+CH_2OH$, $^+CH(CH_3)OH$, page 7]. Another mode of fragmentation in this example involves elimination of the neutral molecule hydrogen chloride, to give a peak at m/z 28 from the positive ion from ethene (ethylene).

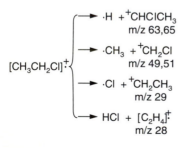

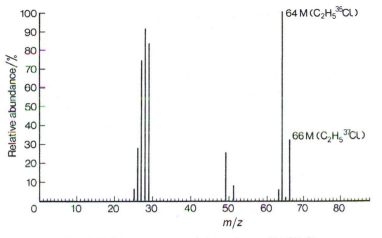

Fig. 1.10 Mass spectrum of choroethane, CH_3CH_2Cl

Diethyl ether (ethoxyethane), $CH_3CH_2OCH_2CH_3$ (Fig. 1.11). The spectrum shows a molecular ion at m/z 74, and an (M-15) peak at m/z 59 which indicates loss of a methyl group: this leaves the fragment $^+CH_2OCH_2CH_3$, in which the oxygen atom is again able to exert a stabilizing influence. The peaks at m/z 45 (M-29) and 29 are consistent with the occurrence of an ethyl group in the molecule; loss of this group gives either $^+CH_2CH_3$ and $^.OC_2H_5$ or $[C_2H_5O]^+$. The peak at m/z 31 is probably from $^+CH_2OH$, arising as shown. The (M-29) peak (m/z 45) probably arises via a similar process. The driving force for these fragmentations is the production of a stable molecule (ethene) with the retention of the positive charge next to oxygen in the remaining cation.

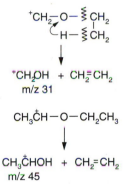

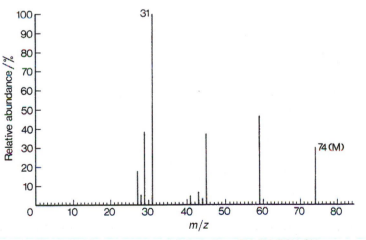

Fig. 1.11 Mass spectrum of diethyl ether, $CH_3CH_2OCH_2CH_3$

Diethylamine, $(C_2H_5)_2NH$ (Fig. 1.12). The odd-numbered molecular ion (m/z 73) confirms a structure with a single nitrogen atom. Fragmentation in this example leads to peaks from $^+CH(CH_3)NHCH_2CH_3$ (m/z 72) and

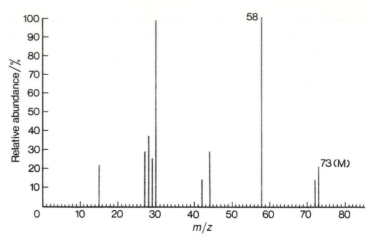

Fig. 1.12 Mass spectrum of diethylamine, $(CH_3CH_2)_2NH$

$^+CH_2NHCH_2CH_3$ [m/z 58 (M-15)], both of which have the positive charge adjacent to the nitrogen atom with its lone pair of electrons. Loss of an ethyl group is indicated by the (M-29) peak at m/z 44; this peak could be from $^+NHCH_2CH_3$, but the rearranged (stabilized) isomer $^+CH(CH_3)NH_2$ seems more likely, as with $^+CH(OH)CH_3$ in the previous example. The peak at m/z 30 is probably from $^+CH_2NH_2$, formed by a similar fragmentation–rearrangement process.

Acetophenone, $C_6H_5COCH_3$ (Fig. 1.13). This mass spectrum shows a molecular ion at m/z 120 and a characteristic (M-15) peak (m/z 105) from loss of a methyl group. Fragmentation at the other side of the carbonyl group also occurs, to give the (M-43) peak at m/z 77.

The detection of a peak at m/z 77 (and, in general, peaks in the 75–77 region) provides very strong evidence for a benzenoid compound. This is also true to a certain extent for the fairly prominent peaks at m/z 50 and 51 formed via the breakdown of the aromatic ring.

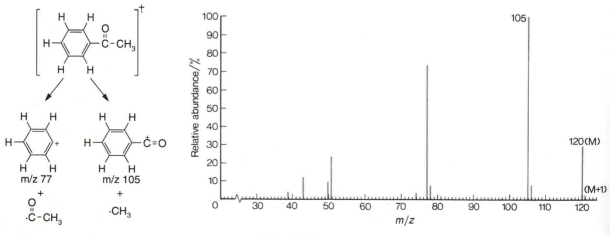

Fig. 1.13 Mass spectrum of acetophenone, $C_6H_5COCH_3$

4-Methylpentan-2-one (methyl isobutyl ketone), $CH_3COCH_2CH(CH_3)_2$ (Fig. 1.14). In this example the peaks at m/z 85 [(M-15)], 43, and 57 [(M-43)] confirm the presence of the $COCH_3$ group. The peak from the $^+CH_2CH(CH_3)_2$ fragment (m/z 57) is not so dominant as that of $^+C(CH_3)_3$ in the isomeric ketone previously considered (Fig. 1.8), which is as expected for the lower stability of the primary [$^+CH_2CH(CH_3)_2$] rather than the tertiary [$^+C(CH_3)_3$] ion. The peak at m/z 58 (M-42) arises via a McLafferty rearrangement, as shown.

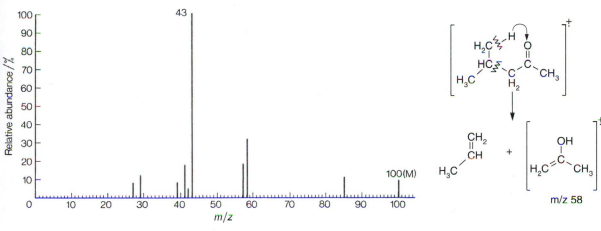

Fig. 1.14 Mass spectrum of 4-methylpentan-2-one, $CH_3COCH_2CH(CH_3)_2$

4-Chlorobenzoic acid, $4\text{-}ClC_6H_4CO_2H$ (Fig. 1.15). This spectrum shows features as expected for a compound with two functional groups. The two peaks at high m/z values, 156 and 158, in the intensity ratio of approximately 3:1 are characteristic of a chlorine-containing compound. The peak at m/z 75 confirms the aromatic nature of the compound. The peaks at m/z 139 and 141, (M-17), again in the ratio 3:1, indicate that fragmentation has taken place (probably with loss of OH) with retention of chlorine in the positive ion. The chlorine is also retained for the fragments at m/z 111/113.

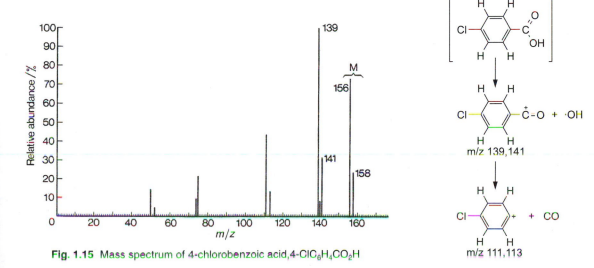

Fig. 1.15 Mass spectrum of 4-chlorobenzoic acid, $4\text{-}ClC_6H_4CO_2H$

The mass spectra of the 2- and 3- isomers of this compound would resemble somewhat that of the 4-isomer and therefore mass spectrometry might not provide unambiguous assignment of a spectrum to one particular isomer. However, structure determination will usually be carried out with a variety of techniques and, for example, the nuclear magnetic resonance and infra-red spectra of the isomers, when examined in conjunction with the mass spectrometry evidence, would normally enable the distinction to be made.

1.5   Worked examples

Figs 1.16–1.18 are the mass spectra of three unknown compounds: the molecular ion (or ions) and the base peaks are denoted, with the ten most intense peaks. With no further information provided, can you identify the compounds?

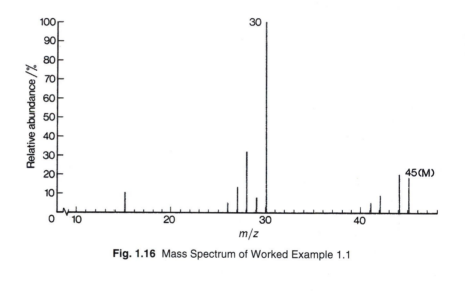

Fig. 1.16 Mass Spectrum of Worked Example 1.1

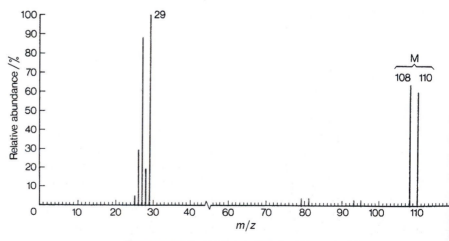

Fig. 1.17 Mass spectrum of Worked Example 1.2

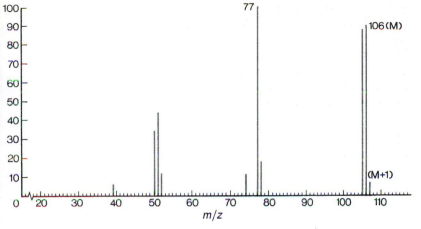

Fig. 1.18 Mass spectrum of Worked Example 1.3

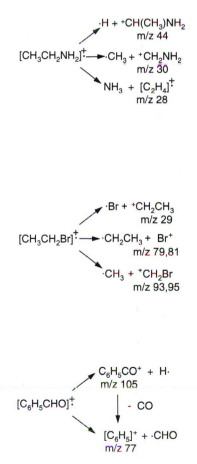

i) Fig. 1.16 is the mass spectrum of ethylamine, $CH_3CH_2NH_2$. The important features to note are the odd-numbered molecular ion, indicating a molecule containing a single nitrogen atom, the intense (M-1) and (M-15) peaks, and the typical 'ethyl-group' peaks at 26–29. All this supports a molecular formula C_2H_7N (rather than say CH_3NO). The breakdown pattern observed derives, in part, from the ability of nitrogen to stabilize an adjacent carbonium ion, and loss of a stable molecule (ammonia in this example) is also observed. Note that the mass spectrum serves to distinguish clearly between this and the isomeric structure $NH(CH_3)_2$: the latter would not show significant ions at 26–29 nor, since $^+NHCH_3$ is not stabilized, at 30 (M-15).

ii) Fig. 1.17 is the mass spectrum of bromoethane, CH_3CH_2Br. The crucial evidence here lies in the 'double' molecular peak, two units of m/z apart, from molecules containing ^{79}Br and ^{81}Br (in almost equal abundance). The peak at m/z 29 strongly suggests an ethyl group, which is confirmed by the peaks from m/z 25 to 28. Other fragments which can be recognized are the bromine ions themselves (m/z 79/81) and also the (M-15) peaks (m/z 93/95). Comparison of this spectrum with that from chloroethane (Fig. 1.10) indicates a greater relative extent of halogen loss in this example, which is consistent with the C–Br bond being weaker than the C–Cl bond.

iii) As judged by the mass spectrum of Fig. 1.18, this compound is aromatic (there is a large peak at m/z 77 and also peaks at 50/51) and it can easily lose one hydrogen atom (to give the large peak at 105; the relative molecular mass is 106). Two possibilities which could lose a fragment of mass 29 to leave the peak at m/z 77 are ethylbenzene and benzaldehyde (benzenecarbaldehyde). However, the former would be expected to give an intense peak at 91 (from phenyl-methyl, $[C_6H_5–CH_2]^+$, see page 8), so the latter structure is preferred. The loss of a hydrogen atom is expected to be particularly facile for benzaldehyde.

1.6   Problems

1.1   Identify the compound whose mass spectrum is shown in Fig. 1.19. Th peak at *m/z* 61 is approximately 2% of the height of that at *m/z* 60.

1.2   Predict the relative intensities and *m/z* values of the molecular ions o different isotopic composition in the mass spectrum of 1,1-dibromo ethane, CH_3CHBr_2.

1.3   Identify the compound (of molecular formula $C_9H_{10}O_2$) whose mas spectrum is shown in Fig. 1.20. How do you account for the formatio of a fragment with *m/z* 108? The infra-red spectrum of this material i on page 40, and the n.m.r. spectrum data is on page 70.

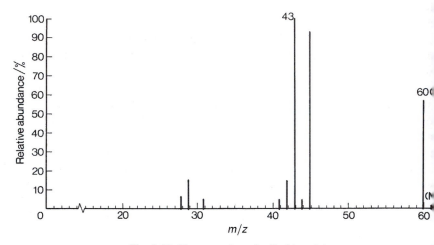

Fig. 1.19 Mass spectrum for Problem 1.1

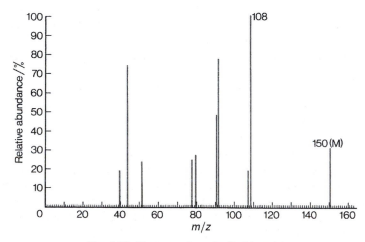

Fig. 1.20 Mass spectrum for Problem 1.3

.7 Further developments and applications

tructural identification

Most mass spectrometers now store the collected spectra in a digital form. This has enabled large collections of spectra to be combined into spectral 'libraries' which can be rapidly examined to search for matches between an nknown compound's spectrum and those in the library. This facility has ramatically assisted the identification of unknown samples.

Mass spectrometry and chromatography

The previous sections demonstrate how the mass spectrometer provides an xtremely sensitive technique for determining the molecular formulae and tructures of a wide variety of organic compounds. Together with an addiional facility it can provide a rapid analysis of complicated mixtures of ifferent compounds.

A small quantity of the mixture to be examined is injected into a gas chromatograph. The chromatograph separates the various volatile components which are partitioned between the carrier gas stream and a liquid phase supported on an inert solid in the column) and these emerge from the chromatoraphic column in the carrier gas stream after different times have elapsed depending on their retention times on the column). Each pure component is ed into the mass spectrometer and its spectrum is recorded with a rapid scan. n this way the *chromatograph* achieves the separation of components (the reas under the chromatographic traces indicate the relative amounts of the eparate components in the mixture) and the *mass spectrometer* provides dditional information which leads to the relative molecular masses and tructures of the separate components. The combination of the two techiques provides a remarkably sensitive and effective method for determining he products, and their relative yields, from a chemical reaction.

Other techniques for volatilization and ionization

Particularly helpful advances include the use of a *fast-atom bombardment* FAB) technique to volatilize and ionize involatile organic samples and the se of *spark excitation* to achieve the necessary volatility for salts.

In **MALDI** spectrometers, the molecule under investigation is embedded n a polymer matrix and subsequently vaporized and ionized by a blast from a aser (Matrix-Assisted Laser Desorption Ionization). In an **Electrospray** mass spectrometer a solution is evaporated, leaving a gaseous molecule in the nlet chamber: for molecules which are already charged (e.g. protonated biological samples) no high-energy treatment is required.

Some techniques which generate ions under more gentle conditions allow much larger biomolecules to be studied very effectively. Where volatilization an be achieved and fragmentation patterns can be observed, very helpful nformation can also be obtained. For example, if a peptide with a possible mino-acid sequence represented as ABCDEF etc. is studied, fragments dentified by their RMM as A, AB, ABC, ABCD etc. allow the whole equence to be defined.

Quantitative applications

We have largely been concerned with the structural aspects of mass spectrometry, with emphasis on the use of wide scan ranges and determination of accurate masses and formulae. However, it should also be noted that smaller 'dedicated' mass spectrometers can give continual monitoring of the level of a chosen substrate or substrates: medical examples include the quantitative analysis of the gases in human lungs during respiration.

A mass spectrometer also allows isotopes to be employed in mechanistic studies: for example, ^{18}O labelling can be used to determine whether hydrolysis of an ester, such as the reaction of H_2O with, e.g. $CH_3C(O)OCH_2CH_3$ proceeds with cleavage of the $CH_3C(O)-O$ or $CH_3C(O)O-CH_2CH_3$ bond.

Other information

Attention can also be focused upon the processes which take place when electrons collide with molecules in the ionization chamber. For example, by studying the appearance of various peaks in a mass spectrum as the energy of the bombarding electrons is increased it is possible to determine the *ionization energy* of the molecule (i.e. the threshold energy at which the collision knocks out an electron) and also the *dissociation enthalpy* (energy) of a bond which is broken in a fragmentation process.

Further reading

1. L. M. Harwood and T. D. W. Claridge, *Introduction to Organic Spectroscopy*, Oxford Chemistry Primers, Oxford University Press, 1997
2. D. H. Williams and I. Fleming, *Spectroscopic Methods in Organic Chemistry*, 5th Edition, McGraw Hill, 1995.

2 Introduction to spectroscopic techniques

Spectroscopic experiments demonstrate that energy can be absorbed or emitted by molecules and atoms in discrete amounts, corresponding to precise changes in energy of the molecule or atom concerned; this is a fundamental part of **quantum theory**. We can measure precisely the amounts of energy involved because when a certain amount of energy is emitted (for example), the energy appears as electromagnetic radiation of a precise frequency (Eqn 2.1).

As will be shown in the following chapters, molecules and atoms are found to absorb only certain energies (frequencies) but not others; similarly, under certain circumstances, discrete frequencies in different parts of the electromagnetic spectrum are emitted.

$$\triangle E = h\nu \qquad (2.1)$$

where $\triangle E$ is the change of energy involved; ν is the frequency of the radiation; h is Planck's constant $(6.626 \times 10^{-34}$ J s$)$

2.1 Electromagnetic radiation: energy, frequency, and wavelength

Visible light is just one of the possible forms of electromagnetic radiation, a fact which is essential to the understanding of spectroscopic and diffraction methods for studying molecular structure. Other forms of this radiation — like X-rays, ultraviolet and infra-red radiation, microwaves, and radiowaves — may be characterized in different ways and may have different applications, but they are essentially the same phenomenon. All forms are characterized by their **frequency** (and hence **wavelength**) and their **energy**.

That different types of radiation possess different energies is not difficult to appreciate (X-rays and ultraviolet radiation can cause much more damage to human tissues than does visible radiation) and it can also be demonstrated convincingly that radiation is wave-like in nature. For example, when light from a single source passes through two narrow parallel slits and then falls on a screen behind the slits, an interference pattern of dark and light lines is produced (Young's slits experiment). In an analogous fashion it can be shown that when a beam of X-rays is incident on a powdered solid, the atoms or ions in the latter behave as a diffraction grating and an interference pattern of maxima and minima is obtained in the reflected X-ray beam. The eye cannot see the interference pattern because X-rays are not visible, but a photographic film is darkened by the reflected beam and reveals the pattern resulting from the interference.

Electromagnetic radiation consists of oscillating electric and magnetic fields which can be transmitted through space. In a vacuum all electromagnetic radiation travels with the same velocity, 3×10^8 m s^{-1}. The oscillations associated with different types of electromagnetic radiation proceed in a wave-like fashion with different *wavelengths* (the distance between successive peaks) and *frequencies* (the number of waves passing a given point per second: the relationship between c, the velocity, λ, the wavelength, and ν, the frequency, is given in Equation 2.2. λ has the units of length (normally m) and ν has the units of time $^{-1}$ [normally s^{-1} or Hz (Hertz)].

$$c = \lambda \times \nu \qquad (2.2)$$

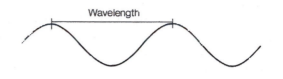

Fig. 2.1 summarizes the approximate ranges of the wavelengths for the different parts of the electromagnetic spectrum. X-rays, for example, are characterized by a very short wavelength (λ is about 10^{-9} m, i.e. 1 nm) and a correspondingly high frequency and energy. Visible light has longer wavelengths, from about 400 nm (which is violet light) to about 750 nm (the red end of the visible part of the spectrum). Radiowaves have much greater wavelengths — of the order of metres.

In some parts of the spectrum the radiation is customarily referred to in terms of its frequency; radiowaves with $\lambda = 100$ m have $\nu = 3 \times 10^6$ Hz, i.e. 3 MHz.

Spectroscopists sometimes also characterize radiation via the reciprocal wavelength, known as the **wavenumber** $\bar{\nu}$; like frequency, wavenumber is directly proportional to energy. Because of its almost universal usage for infra-red radiation the wavenumber will be retained here in that context. The unit customarily employed is cm^{-1}, i.e. waves per centimetre; if $\lambda = 10^{-6}$ m, the wavenumber, $\bar{\nu}$, is 10^4 cm^{-1}.

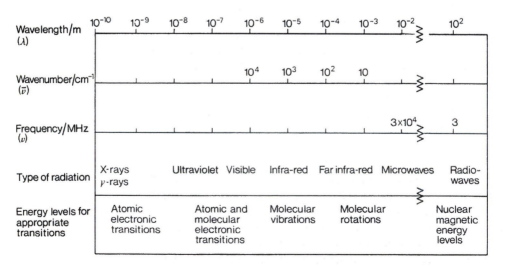

Fig. 2.1 The electromagnetic spectrum

2.2   Atomic and molecular energy levels

Atoms

Our knowledge of the well-defined energy levels for electrons in atoms orig-
inates in spectroscopic observations on the radiation emitted when atoms or
ions become excited. Think of the simple flame test for alkali metals; for
example, a small quantity of a sodium salt, when placed in a flame, appears
yellow. This is because sodium atoms are formed and, at the high temperature of the flame, these become excited by absorption of energy; this energy
is subsequently emitted as radiation in the visible region. The experiment can
be carried out more precisely by analysing the emitted radiation in terms of
the frequencies (and hence wavelengths) present; the radiation is incident on
a prism which disperses radiation of different wavelengths in different directions (like Newton's experiment to produce all the colours in the visible spectrum from 'white' light). The instrument which is used to investigate this is
called a spectroscope (hence the term *spectroscopy* for studies of this kind)
and the resulting radiation (not all of which will be in the visible region) can
be detected with a photographic film.

It can then be demonstrated that from excited atoms only certain wavelengths are emitted: the resulting plot of wavelength contains several discrete
values or 'lines' — these are typical line-spectra. In the example of the
sodium flame (and, in terms of a useful application, the sodium-vapour street
lamp), the orange colour derives mainly from the intense emission with
$\lambda = 589$ nm. Line-spectra can be used to identify different elements present
(e.g. in a vaporized sample) and also, under certain circumstances, to determine the amount of a particular material present (from the intensity of the
radiation associated with a particular transition).

The spectra can also be interpreted in terms of the energy levels associated
with electrons in various shells (or orbitals) in the atom concerned. For
example, the emission with $\lambda = 589$ nm for the sodium atom corresponds to
the energy emitted when an electron which has been promoted to a $3p$ orbital
returns to the lower energy $3s$ orbital. The energy change can be larger if
'inner' electrons are involved; for example, bombardment of copper with
electrons leads to the ejection of an inner $(1s)$ electron and the resulting transition of an outer electron from a higher energy level down to this vacancy, for
example from $2p$ to $1s$, leads to the emission of X-rays ($\lambda = 0.15$ nm in this
case).

Absorption of exact amounts of energy — a process leading to a transition
from a lower *to* a higher energy level — is also possible. A clear example
here is provided by the **Fraunhofer** lines: in the otherwise continuous spectrum of visible radiation from the sun, several wavelengths are missing.
These correspond exactly to lines in the emission spectra of various atoms,
including hydrogen. The explanation is that the continuous radiation from the
sun passes through its atmosphere which contains these atoms; the atoms
absorb energy of discrete wavelengths which are then missing from the radiation reaching the earth. This provides a method for detecting gases in the
atmosphere of the sun. Atomic Absorption Spectroscopy (AAS) similarly
provides a useful method for determining the concentration of, for example,
trace metals in aqueous solutions.

Molecules

It can also be shown that *molecules* have certain exact energies, and absorption of the relevant frequencies from incident radiation raises molecules from lower to higher levels: absorptions may be detected in three different areas of the electromagnetic spectrum.

First, the **electrons** in molecules occupy molecular orbitals with precise energy levels (cf. electrons in atomic orbitals). Transitions from lower, filled orbitals to upper (higher energy) empty orbitals usually involve absorption of radiation in the *ultraviolet* (u.v.) and *visible* parts of the spectrum. This is the basis of electronic absorption spectroscopy, described in Chapter 4.

Much smaller quantities of energy are associated with changes in the **rotational** energy of a molecule (which is allowed only certain well-defined values) and in its **vibrational** energy (which is also quantized). This second area of interest is concerned with precise energy absorption in the *infra-red* (i.r.) part of the spectrum. How the measurement and interpretation of these energy changes in molecules leads to structural information is described in the next chapter.

The energy changes associated with certain **nuclei** in magnetic fields are smaller still and occur in the radio-frequency region. This is the basis of *nuclear magnetic resonance* spectroscopy, n.m.r. (Chapter 5).

Differences in the magnitudes of the energy changes involved dictate the necessity for different instrumental arrangements for u.v. (and visible), i.r. and n.m.r. spectroscopy. Understanding why a particular wavelength of radiation is emitted (or absorbed) then leads to detailed information about the molecules involved and provides a key to the investigation of their structure.

Further Reading

1. L. M. Harwood and T. D. W. Claridge, *Introduction to Organic Spectroscopy*, Oxford Chemistry Primers, Oxford University Press, 1997.
2. C. N. Banwell and E. M. McCash, *Fundamentals of Molecular Spectroscopy*, 4th Edition McGraw Hill, 1994.

3   Infra-red spectroscopy

Infra-red (i.r.) radiation is the term used to describe electromagnetic radiation with frequencies and energies somewhat lower than those associated with visible light; it is emitted as a range of frequencies from a heated object (sometimes together with visible radiation). When a beam of i.r. is incident upon a collection of certain molecules, absorption of discrete frequencies by the molecules takes place, corresponding to the absorption of well-defined amounts of *energy* from the range of energies in the radiation. This is the basis of i.r. absorption spectroscopy.

Two main applications of i.r. spectroscopy provide important structural information about molecules. The first is the study of simple molecules (diatomic and triatomic) in the gas phase; the exact amounts of energy absorbed from the i.r. radiation are related to increases in the rotational and vibrational energy of the molecules. It is possible to determine *bond lengths* and also *force constants* (a measure of the resistance to stretching).

The second application of i.r. involves the recognition of the *structures* of more complicated molecules from their characteristic absorptions. As with mass spectrometry and n.m.r. spectroscopy (Chapter 5), i.r. can be used to indicate the nature of the functional groups in a molecule, and, by comparison with spectra from known compounds, to aid identification of an unknown material.

3.1   Pure rotation i.r. spectra of small molecules

For gaseous diatomic molecules, an absorption of i.r. radiation is only possible if the molecule has a **dipole moment**. This occurs when the two atoms are chemically different (e.g. HF), such that an unequal sharing of electrons leads to an asymmetric distribution of electron density.

The molecule is rotating, both about an axis along the bond and also about an axis through the centre of gravity and perpendicular to the bond (Fig. 3.1). The latter motion gives rise to a fluctuating electric field which enables this type of molecule to interact with the fluctuating electric field of the incoming radiation and hence, by absorbing energy from the radiation, to increase its rotational energy. Molecules without dipoles rotate in similar fashion, but do not interact in this way with incident radiation.

Fig. 3.2 shows the spectrum obtained when radiation with wavenumber ($\bar{v}$) in the 0–300 cm^{-1} range is incident upon a gaseous sample of HF; there are several absorptions of energy, recognized by the downward peaks, at characteristic wavenumbers (i.e. at certain exact energies), and the separation between the lines is approximately constant (40.5 cm^{-1}). Similar spectra, but with different separations, are obtained for other heteronuclear diatomic molecules (for HCl, the spacing is 20.7 cm^{-1}).

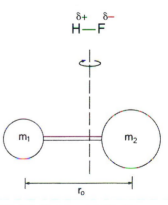

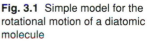

Fig. 3.1 Simple model for the rotational motion of a diatomic molecule

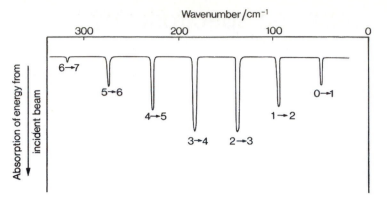

Fig. 3.2 Rotational i.r. absorption spectrum (in the range $\bar{v}$ 0–300 cm⁻¹) for gaseous hydrogen fluoride, HF

$$E_{rot} = \frac{h^2 J(J+1)}{8\pi^2 I} \qquad (3.1)$$

The explanation for the appearance of the spectrum is as follows. For a molecule which consists of two masses m_1 and m_2 and a bond length r_0, the only allowed values for the rotational energy (E_{rot}) are given by eqn 3.1 where h is Planck's constant (6.626×10^{-34} J s), I is the *moment of inertia* for rotation about the axis indicated in Fig. 3.1 (given by $[m_1 m_2/(m_1 + m_2)]r_0^2$). J is a *quantum number*, with the allowed values of 0,1,2,3… Only certain rotational energy levels (E_{rot}) for the molecule are allowed, because the quantum number is restricted to certain values: the energy levels are indicated in Fig. 3.3. Each molecule at any moment must be in one of these energy levels, and since the energy separations between the levels are small, all the levels indicated will be fairly well populated (the distribution of molecules amongst energy levels is discussed later).

A molecule can increase its rotational energy by being promoted from one level to the next level (i.e. $\Delta J = 1$) if *exactly* the correct amount of energy is incident upon it. Thus, those specific wavenumbers whose energies correspond to the changes of energy from $J = 0$ to $J = 1$ (by molecules in the $J = 0$ level), or from $J = 1$ to $J = 2$ (by molecules in the $J = 1$ level), etc., are absorbed from the incoming radiation.

For an allowed transition, $J \rightarrow J'$ the energy change involved can be written:

$$\Delta E_{rot} = \frac{h^2}{8\pi^2 I}\left[J'(J'+1) - J(J+1)\right]$$

and therefore, since $\Delta J = 1$ (i.e. $J' = J+1$) then:

$$\Delta E_{rot} = 2J'\left(\frac{h^2}{8\pi^2 I}\right) \qquad (3.2)$$

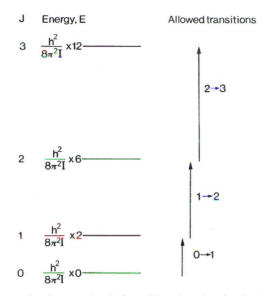

J   Energy, E   Allowed transitions

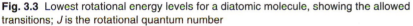

Fig. 3.3 Lowest rotational energy levels for a diatomic molecule, showing the allowed transitions; J is the rotational quantum number

where J' is the quantum number for the upper state. The energies of the transitions from the lowest levels are as follows:

Transition	Energy change
$0 \rightarrow 1$	$\left(\dfrac{h^2}{8\pi^2 I}\right) \times 2$
$1 \rightarrow 2$	$\left(\dfrac{h^2}{8\pi^2 I}\right) \times 4$
$2 \rightarrow 3$	$\left(\dfrac{h^2}{8\pi^2 I}\right) \times 6$

This means that transitions from successive energy levels to the levels above them are associated with energy changes (ΔE) which have steadily increasing values (an increment of $2h^2/8\pi^2 I$). This is exactly what is observed in the spectrum: thus, for HF, molecules undergoing the $0 \rightarrow 1$ transition absorb energy corresponding to a wavenumber of about 40 cm^{-1}; for molecules undergoing the $1 \rightarrow 2$ transition $\Delta \bar{\nu}$ is about 80 cm^{-1}; for the $2 \rightarrow 3$ transition $\Delta \bar{\nu}$ is about 120 cm^{-1}, and so on.

Calculation of the bond length of HF

The difference ($\Delta \bar{\nu}$) between two successive rotational lines is 40.5 cm^{-1}, i.e. $\Delta \nu = 1.22 \times 10^{12}$ Hz. Since $E = h\nu$, we can convert $\Delta \nu$ to the appropriate energy difference (by multiplication by Planck's constant). The result must equal $(2h^2/8\pi^2 I)$, as shown above. From this calculation I and r_0 can be obtained.

$$\Delta E = h \Delta v = \frac{2h^2}{8\pi^2 I}$$

$$I = \frac{h}{4\pi^2 \Delta v} = \frac{6.626 \times 10^{-34}}{4\pi^2 \times 1.22 \times 10^{12}}$$

$$= 1.376 \times 10^{-47} \text{ kg m}^2 \text{ (or J s}^2\text{)}$$

Now, since

$$I = \left(\frac{m_1 m_2}{m_1 + m_2} \right) r_0^2 \tag{3.3}$$

where m_1 is the mass of the hydrogen atom, and m_2 is the mass of the fluorine atom, so that $m_1 = \dfrac{1}{10^3 \times L}$ kg, $m_2 = \dfrac{19}{10^3 \times L}$ kg, where L is the Avogadro Constant (6.023×10^{23} mol^{-1}), then,

$$r_0^2 = 1.376 \times 10^{-47} \times \frac{20}{19} \times 6.023 \times 10^{26} = 0.872 \times 10^{-20} \text{ m}^2$$

$$r_0 = 0.93 \times 10^{-10} \text{ m} = 0.093 \text{ nm}$$

Table 3.1 Bond lengths/nm for diatomic molecules determined from rotational spectra

HF	0.093
HCl	0.127
HBr	0.141
HI	0.160

In this way the bond lengths (r_0) of a variety of heteronuclear diatomic molecules can be measured (see Table 3.1).

In practice, the experiments are most accurately carried out with what is known as a **microwave** (rather than a far infra-red) source of radiation (the essential theory and the range of wavelengths employed is the same in the two cases, although the experimental arrangements differ somewhat). Bond lengths can then be estimated to within about 0.0005 nm. It should also be noted here that the molecules are not rigid but are actually vibrating (see later) so that the bond length measured is an *average* value. At higher rotational energy levels the average bond length also shows centrifugal distortion, and the spacings change slightly.

This interpretation of the rotational spectrum of a diatomic molecule can be tested by employing an isotopic substitution method. The bond length in a molecule is an *electronic* property and should not be affected by isotopic substitution, so that ^{1}HF and ^{2}HF (deuterium fluoride) should have the same bond length. The spacing in the rotational i.r. spectrum for ^{2}HF is found to be approximately half that for ^{1}HF, which is just as expected from eqns 3.2 and 3.3 (you are encouraged to check this calculation).

Triatomic molecules

Complications arise when a linear triatomic molecule (e.g. HCN) is investigated. A spectrum can be observed, since the molecule has a dipole moment, and the discrete energy absorptions correspond to quantized changes in the energy of rotation about an axis through the centre of gravity, perpendicular to the bonds. However, the spacing in the spectrum leads only to a single moment of inertia, and this value cannot be used to derive *both* bond lengths, (r_{CH} and r_{CN} for this example). The problem can be solved, however, with the help of isotopic substitution; the spacing of the energy levels, and hence the moment of inertia is measured for ^{2}HCN, as well as for ^{1}HCN. As the bond lengths are unaltered by isotopic substitution, there is enough information (two moments of inertia, known masses) for both r_{CH} and r_{CN} to be determined.

For more complicated molecules, an analysis to give bond lengths and angles may be possible if there is some degree of symmetry (as well as a dipole moment). For example, from a study of trichloromethane (chloroform, $CHCl_3$) the moment of inertia about an axis perpendicular to the C−H bond can be determined (rotation about an axis along this bond leads to no change in dipole moment). Then, via experiments on isotopically labelled derivatives, these details of the structure can be established.

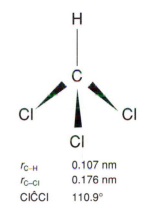

r_{C-H}	0.107 nm
r_{C-Cl}	0.176 nm
ClĈCl	110.9°

3.2 Vibration–rotation i.r spectra of small molecules

Many molecules show, in addition to the absorptions in the microwave (far i.r.) region of the spectrum, characteristic absorptions in a higher-energy part of the i.r. region. Most commercial spectrometers are designed to operate in this range, with gaseous liquid and solid samples. In addition to the measurement of bond lengths for certain molecules, considerable extra information can then be obtained.

Infra-red spectrometer

Fig. 3.4 illustrates the essential features of a typical instrument. The radiation is emitted from a heated filament as a continuous range of frequencies (and hence wavelengths and wavenumbers) in the i.r. region. This radiation is then passed through the sample, which is contained in a cell with a path length, for a gas, of several centimetres, or, for a liquid sample, of up to 10^{-2} cm. The resulting radiation, from which some of the radiation at certain frequencies will have been absorbed by molecules in the sample, is then passed through a system of slits and mirrors to emerge as a collimated beam. A prism or

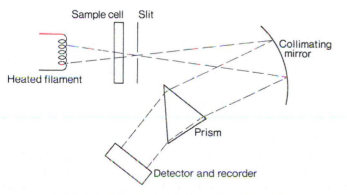

Fig. 3.4 Essential features of an i.r. spectrometer

grating disperses the beam into components at different wavelengths (just as a prism splits up a beam of visible white light into different colours) and depending on the orientation of the prism, radiation of separate wavelengths reaches the detector thermocouple. The detector monitors the radiation transmitted at different wavelengths and converts the radiant energy into an electrical signal. An automatic scan of frequency or wavenumber, against energy either absorbed or transmitted, is easily achieved. The prism and sample holders have to be transparent to i.r. radiation and are prepared from suitable inorganic salts (NaCl, KBr, CaF_2, for example).

The spectrum from hydrogen chloride

Fig. 3.5 shows the absorption spectrum of gaseous HCl in the wavenumber region 2600–3100 cm^{-1}. The most important features to note are the regular spacings of about 20 cm^{-1} between adjacent lines and the fact that the spectrum is centred at about 2890 cm^{-1}. Lines from the isotopically different species $^1H^{35}Cl$ and $^1H^{37}Cl$ are not seen separately in this spectrum (see page 31).

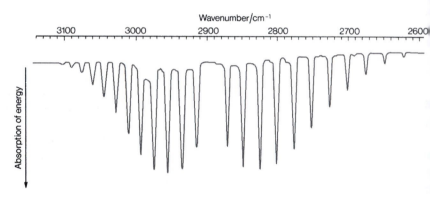

Fig. 3.5 Vibration rotation i.r. spectrum ($\bar{v}$ 2600–3100 cm^{-1}) of gaseous hydrogen chloride, HCl

Analysis of the spectrum

The behaviour of hydrogen chloride can be understood in terms of the molecule having **vibrational** energy (again, in well-defined amounts, or **quanta**) as well as rotational energy. For molecules with a dipole moment (e.g. HCl) this vibration allows interaction with the incident radiation, and hence energy can be absorbed: radiation of the appropriate energy (greater than that required for changes in the rotational energy) raises the molecule from its lowest *vibrational* energy state to the first excited vibrational state. The rotational energy can also change, which accounts for the many lines observed. This is now explained in more detail.

The quantum theory predicts that only certain vibrational energies E_{vib} (eqn 3.4) are allowed, where v is a quantum number, with possible values 0, 1, 2 etc. and v_0 is called the **fundamental frequency.**

$$E_{vib} = (v + \tfrac{1}{2})hv_0 \qquad (3.4)$$

The lowest two energy levels, with $\mathbf{v} = 0$ and $\mathbf{v} = 1$, will have $E_{vib} = \frac{1}{2}(h\nu_0)$ and $E_{vib} = \frac{3}{2}(h\nu_0)$ respectively, so that the difference between them (i.e. the energy of the $\nu_0 \rightarrow \nu_1$ transition) is $h\nu_0$: the appropriate frequency associated with this energy change is the fundamental frequency (ν_0). It should also be noted that even in the ground vibrational state the molecule has vibrational energy, $\frac{1}{2}(h\nu_0)$; this is called the zero point energy.

The molecule is rotating and vibrating simultaneously, and the *total* energy associated with these motions is the sum of the separate rotational and vibrational energies previously referred to (see eqn 3.5). This means that there are many possible energy levels: a molecule in a certain vibrational energy level can still have any one of the possible rotational energy levels referred to earlier.

$$E_{vib} + E_{rot} = (\mathbf{v} + \frac{1}{2})h\nu_0 \qquad (3.5)$$
$$+ \frac{h^2 J(J+1)}{8\pi^2 I}$$

Some of the allowed energy levels are illustrated in Fig. 3.6: note that the vibrational energy levels are much more widely spaced than those for rotation (i.e. *ca.* 3000 cm^{-1} compared with 20 cm^{-1} — the vibrational energy separation in the figure is not to scale).

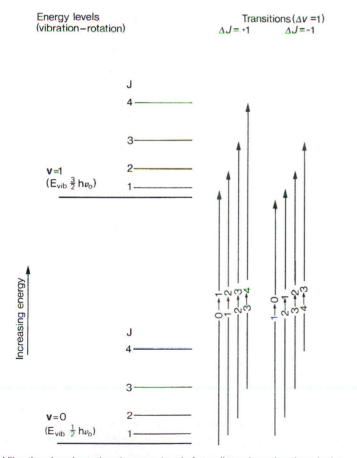

Fig. 3.6 Vibrational and rotational energy levels for a diatomic molecule: $\mathbf{v}$ is the vibrational quantum number and J is the rotational quantum number

The allowed changes in vibrational and rotational energy levels are governed by selection rules for Δv and ΔJ.

When a molecule absorbs energy it can either change its J value (as discussed in Section 3.1) or it can change its vibrational *and* rotational levels. The particular transitions of the latter type which it is allowed to undergo are limited to those for which $\Delta v = +1$ (e.g. from $v = 0$ to $v = 1$, $v = 1$ to $v = 2$) *and* $\Delta J = \pm 1$. For example, a molecule in the energy level $v = 0$, $J = 1$ can absorb energy to be promoted either to the level $v = 1$, $J = 2$, *or* to $v = 1$, $J = 0$. When all the possibilities are considered it is found that the spectrum should consist of two series of lines — from a given J value in $v = 0$ to $(J+1)$ in $v = 1$, and from J in $v = 0$ to $(J - 1)$ in $v = 1$. The two series will have increasing energy (and hence increasing wavenumber) and decreasing energy (and wavenumber), respectively, as seen in Fig. 3.6 You are encouraged to identify which peaks are associated with individual transitions (0–1, etc), to measure the relative ΔE values for the transitions indicated, and to check that these account for the features of a typical spectrum (Fig. 3.5).

The whole spectrum is centred about the transition $(v = 0) \rightarrow (v = 1)$ which has no associated change in J (i.e. $\Delta J = 0$). For the molecule HCl, this transition is forbidden, and does not take place, but the corresponding value of the energy change associated only with vibration (i.e. as if $\Delta J = 0$) can be obtained from the middle of the spectrum.

In the analysis, only transitions from $v = 0$ to $v = 1$ are indicated: this is a very reasonable approximation, since most of the molecules present will be in their lowest vibrational level rather than in higher levels ($v = 1$, 2 etc.). In contrast, as indicated by the rotational spectrum (e.g. Fig. 3.2), there are molecules in a wide range of rotational levels. This situation arises because the energy increment associated with the allowed increase of vibrational energy is much larger than that for an increase of rotational energy, so that molecules have only a small probability of being in an excited vibrational state.

The profile of the peak intensities in the two branches of the vibration–rotation spectrum (page 28) reflects the Boltzmann distribution of molecules within different rotational energy levels (see page 56).

The analysis also explains why the individual absorptions in the rotation-vibration spectrum are approximately equally spaced. This separation is identical to that obtained for the simple rotational spectrum, so that bond lengths can be obtained as described previously (page 26). It is helpful also to envisage this set of absorptions as centred at 2890 cm^{-1} (corresponding to the $v_0 \rightarrow v_1$ transition and to the fundamental frequency v_0). For compounds in solution the rotational 'fine structure' becomes blurred and a single broad peak centred on v_0 is observed.

Interpretation of the fundamental frequency, v_0

The motion of the vibrating atoms of a diatomic molecule is closely analogous to the simple harmonic motion of two masses attached to each other by a spring (Fig. 3.7). Both systems obey Hooke's law; that is, the restoring force when the masses are stretched or compressed away from the equilibrium position is proportional to the extent of displacement from that position. For the resultant simple harmonic motion of the spring it can be shown that the frequency of oscillation, v, is given by equation 3.6. where μ is the **reduced mass**, $m_1 m_2 / (m_1 + m_2)$, and k is the force constant of the spring (a measure of its resistance to stretching). Similarly, the behaviour of chemical bonds can be interpreted in terms of a fundamental frequency, v_0, given by equation (3.6), where μ is the reduced mass of the two connected atoms, and, by analogy, k is the force constant of the bond. This concept is helpful in understanding vibrational spectra.

$$v = \frac{1}{2\pi}\sqrt{\frac{k}{\mu}} \qquad (3.6)$$

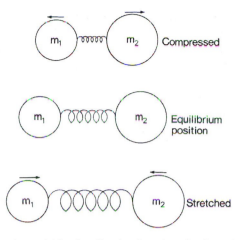

Fig. 3.7 Simple model for the vibrational motion of a diatomic molecule

For $H^{35}Cl$, the absorption at wavenumber 2890 cm^{-1} in the vibrational spectrum is equivalent to a fundamental frequency, ν_0, of 8.67×10^{13} Hz. This leads to a value for the force constant k, of 4.8×10^2 N m^{-1} or kg s^{-2} (4.8×10^5 dynes cm^{-1}).

The equations for rotational and vibrational energy levels indicate that the lines for $H^{35}Cl$ and $H^{37}Cl$ should be almost superimposed (as observed) because μ, that is $m_1 m_2/(m_1 + m_2)$, is very similar for the two molecules.

However, the equations indicate that this similarity does not apply if we compare HCl and DCl. Thus the predicted spectrum from the latter (either chlorine isotope) will have quite different energies for its rotational and vibrational transitions when compared with the former. The force constant, an electronic property, is the same for HCl and DCl (with either chlorine isotope in each case) so that ν_0(HCl) and ν_0(DCl) should be related, as indicated by eqn 3.6, by a factor given by their different $\sqrt{\mu}$ values: the experimental spectrum (Fig. 3.8) confirms the expected ratio of $\sqrt{2}$ for ν_0 (HCl):ν_0 (DCl), with ν_0 for DCl *ca.* 2090 cm^{-1} (remember that wavenumber is proportional to frequency). The spectrum also confirms the expected smaller rotational splitting for DCl; you may like to confirm that a factor of 2 (as observed) is predicted by the theory given here. Further, the separation of the spectrum from ^{2}HCl (Fig. 3.8) into absorptions from $^2H^{35}Cl$ and $^2H^{37}Cl$, with slightly different ν_0, is also as expected from eqn 3.6: the relative intensities of the two sets of peaks are in the ratio 3:1, as expected. (Note the separation is smaller for $^1H^{35}Cl$ and $^1H^{37}Cl$ so that the separate absorbances cannot be resolved in Fig. 3.5).

The spectra of other heteronuclear diatomic molecules can also be analysed to give appropriate force constants. As might be expected, there is a correlation between force constants and bond enthalpies (energies) which is apparent, for example, in the values for the series of hydrogen halides (Table 3.2): the trend in force constants also parallels the changes in bond length (see Table 3.1, page 26). Similarly, the strong triple bond in carbon monoxide (bond enthalpy 1075 kJ mol^{-1}) has a correspondingly large force constant (18.4×10^2 N m^{-1}).

Table 3.2

	Force Constant/ N m^{-1}	Bond Enthalpy/ kJ mol^{-1}
HF	9.7×10^2	562
HCl	4.8×10^2	434
HBr	4.1×10^2	366
HI	3.2×10^2	299

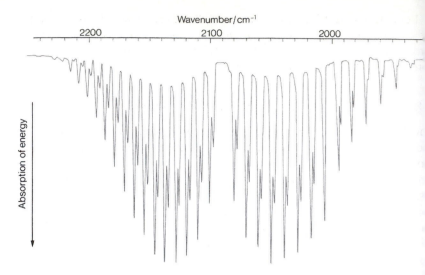

Fig. 3.8 Vibration rotation i.r. spectrum ($\bar{v}$ 1900–2300 cm⁻¹) of gaseous deuterium chloride, ^{2}HCl (DCl)

Another branch of spectroscopy known as Raman Spectroscopy can be employed to calculate bond lengths and force constants for homonuclear diatomic molecules (e.g. H_2, N_2), though it will not be discussed further here.

Triatomic molecules

For molecules more complicated than the diatomic molecules considered so far there are several possible ways in which the bonds can vibrate (these are called vibrational **modes**) and each vibration has an associated fundamental frequency. For example, the fundamental modes for the linear molecule CO_2 are illustrated in Fig. 3.9. Of these, v_1 (the *symmetric stretching* mode) does not involve a dipole moment change and is 'inactive' in the infra-red region, there being no corresponding absorption of energy. However, absorptions are observed for v_2 (667 cm⁻¹) and v_3 (2349 cm⁻¹) because these modes of vibration involve dipole moment changes. These are called *bending* (v_2) and *asymmetric stretching* (v_3) modes, respectively, and the numerical values indicate (as might be expected) that less energy is involved in bending than in stretching.

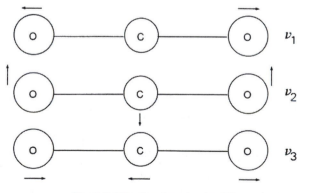

Fig. 3.9 Vibrational modes for CO_2

For a non-linear triatomic molecule (e.g. H_2O) the three vibrational modes e all infra-red active since each involves a dipole moment change. Since ree absorptions are detected in the vibrational i.r. spectrum of SO_2, it can be ncluded that this molecule (like H_2O) is not linear.

3 I.r. spectroscopy of organic molecules

ne i.r. absorption spectrum gets very rapidly more complicated as we ogress from diatomic molecules to tri-atomic, tetra-atomic, and larger olecules, because of the increase in the number of possible vibrations. nly in fairly simple cases can a full analysis be carried out, but neverthe-ss it is often possible to obtain useful information about the structure of mplex molecules. The compound under investigation is examined, where ossible, as a liquid sample. This can be done, for example, for a neat liquid ' squeezing a few drops between two KBr discs (transparent to i.r. radia-on). Alternatively, the compound to be studied may be dissolved in a suit-le solvent (e.g. $CHCl_3$, CCl_4); although peaks characteristic of the lvent's i.r. absorptions will be observed, these can easily be recognized d allowance made for them. (In a double-beam spectrometer, peaks from e pure solvent are recorded simultaneously and subtracted automatically.) solid compound may be ground up to form a 'mull' with paraffin oil fore being placed between the KBr discs or ground with KBr and pressed to a disc.

The spectra are plots of **transmittance** of energy at a given wavelength, ther than absorption of energy. Transmittance and absorption are related: a ʳge transmittance implies little absorption, and vice versa.

Organic compounds show spectra in which many peaks are spread over e wide scan-range customarily employed (4000–600 cm^{-1}). Each peak is sociated with a particular vibration (or a combination of these) — the rota-onal fine structure is smeared out for molecules in the liquid phase because otation of the molecules is not free, as it is in a gas; each peak is really an nvelope' of all rotational lines. The complexity of the spectra (see, for ample, the i.r. spectrum of propanone, Fig. 3.10) reflects the large number fundamental vibrations, which depends on the complexity of the molecule

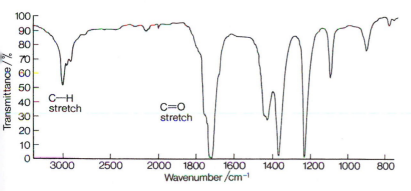

Fig. 3.10 I.r. spectrum of propanone, $(CH_3)_2CO$ (liquid film)

(i.e. its number of bonds). Fortunately, to a certain extent certain absorption can be associated with stretching or bending vibrations of particular bonds (or sometimes groups) in a molecule: for example, the absorptions at *ca.* 3000 and 1700 cm^{-1} in the spectrum of propanone are typical of the stretching modes of C$-$H and C$=$O groups, respectively, in organic compounds. In this way the spectra allow recognition of the types of functional group present in an organic molecule.

Other absorptions are observed which are characteristic of the molecule as a whole, these depending on interactions between different groups. Further, there are complications which arise because absorptions can occur at overtones (harmonics) of other frequencies or at combinations ($\nu_1 + \nu_2$) and differences ($\nu_1 - \nu_2$) of other frequencies. These all add to the complexity but help provide the equivalent of an unambiguous fingerprint for any particular molecule.

Dependence of the i.r. spectrum on molecular structure

Typical absorptions indicated diagrammatically in Fig. 3.11 can be of considerable assistance in the determination of an unknown compound's structure. The important features of the figure can be helpfully interpreted in terms of the dependence of a characteristic bond-stretching frequency (and hence wavenumber) on the *force constant* for that bond and on the *masses* of the atoms which are joined by the bond.

For example, the lowering of the wavenumber (and hence energy) for stretching in the series *triple bond > double bond > single bond* [for C$-$C and C$-$N bonds and, if carbon monoxide is included (C$\equiv$O, ν_0 2146 cm^{-1}), for C$-$O bonds], is as expected from the appropriate bond enthalpies (bond strengths) and hence force constants (see eqn 3.6). Further, variations in the reduced mass (μ) of the atoms forming the bond (see eqn 3.6) also have predictable effects, notably high wavenumber (and hence frequency) for C$-$H, N$-$H, and O$-$H stretches (these bonds all have low values of μ). Lastly, since bending involves less energy than stretching, absorptions for the former occur at lower wavenumbers than stretching modes involving the same bond.

Absorptions tend to be particularly prominent when a bond with large dipole moment is involved (e.g. C$-$O).

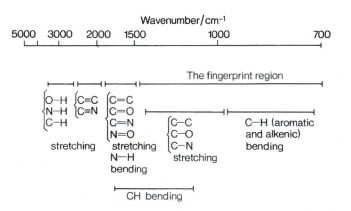

Fig. 3.11 Typical areas for absorptions in the i.r. spectra of organic molecules

One of the particular attractions of i.r. spectroscopy is that a more detailed investigation of the positions of absorption of a bond (e.g. C—H) in a variety of different molecules shows that the exact wavenumber does depend to a small but significant extent on the environment of that bond in a molecule (cf. chemical shifts in n.m.r. spectra; Chapter 5) and this can prove diagnostically extremely valuable. Some characteristic absorptions which illustrate the variation are listed in Table 3.3 and these are now discussed in more detail.

(i) C—H *bonds*. For example, methyl (CH_3) and methene (CH_2) groups usually exhibit C—H stretching modes in the range 2950–2850 cm^{-1} (sometimes these appear with splittings owing to interaction between the different C—H bonds) and also bending modes at around 1450 cm^{-1}. The absorptions for the C—H stretch in propanone at just below 3000 cm^{-1} and the bending modes at *ca.* 1400 cm^{-1} are clearly visible in Fig. 3.10.

In contrast, an alkanal C—H stretch usually appears between 2700 and 2900 cm^{-1}, and other C—H stretching absorptions are as follows: alkynes (C≡CH), 3300 cm^{-1}; alkenes (C=CH$_2$), 3095–3075 cm^{-1}; arenes 3040–3010 cm^{-1}. Out-of-plane bending for arene and alkenic hydrogen atoms often gives characteristic absorption bands at 900–650 and 990–890 cm^{-1}, respectively.

Table 3.3 Characteristic infra-red absorptions for a variety of organic molecules*

Molecule or group	Vibration type	Wavenumber/cm^{-1}
Alkyl group (CH$_3$,CH$_2$, CH)	C—H stretch	2960–2850
	C—H bend	1460–1370
Alkanal (CHO)	C—H stretch	2900–2700
Alkyne (C≡CH)	C—H stretch	3300–3270
Alkene (C=CH$_2$)	C—H stretch	3095–3075
	C—H bend	990–890†
Arene	C—H stretch	3040–3010
	C—H bend: in-plane	1300–1000
	C—H bend:out-of-plane	900–650†
Alkanol (OH)	O—H stretch	3650–3590◊
	C—O stretch	1200–1050
Amine, amide (NH$_2$)	N—H stretch	3500–3300◊
Aliphatic ketone (R$_2$CO)	C=O stretch	1740–1700
Aliphatic alkanal (RCHO)	C=O stretch	1740–1720
Aromatic ketone (Ar$_2$CO)	C=O stretch	1700–1680
Alkanoic acid (RCO$_2$H)	C=O stretch	1725–1700
Alkanoyl chloride (RCOCl)	C=O stretch	1815–1790
Alkanoate ester (RCO$_2$R′)	C=O stretch	1750–1730
	C—O stretch	1300–1050
Alkoxy (ether) R$_2$O	C—O stretch	1150–1070

Some of these peaks may be split into several components: see above.
Characteristic variations occur with different substitution patterns.
◊ These may be drastically affected by hydrogen-bonding.

The characteristic arene C—H stretching vibrations (>3000 cm⁻¹) and out of-plane bending vibration (680 cm⁻¹) are apparent in the i.r. spectrum o benzene (Fig. 3.12). Other prominent absorptions are from in-plane C—C bending (1040 cm⁻¹) and C—C stretching (1480 cm⁻¹). For more complicatec benzene derivatives, it is sometimes possible to determine the ring substitu tion pattern from the modifications produced in the i.r. spectrum.

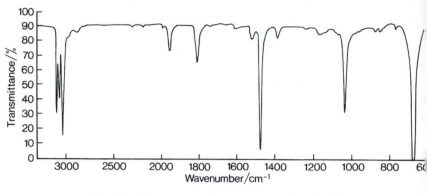

Fig. 3.12 I.r. spectrum of benzene, C_6H_6 (liquid film)

(ii) O—H, C—O *bonds*. Hydroxyl groups (OH) generally give a strong peak (from the stretching vibration) in the 3650–3590 cm⁻¹ region; if the group is hydrogen-bonded there is usually a broadening and a shift towards lower wavenumbers. For example, Fig. 3.13 shows the i.r. spectrum of ethanol showing clearly the C—H and O—H stretches (2900 and 3300 cm⁻¹, respectively) and absorption from aliphatic C—H bending (1400 cm⁻¹) anc C—O stretching 1050 cm⁻¹, see also section (iv). From the shape and position of the O—H peak it can be concluded that this group is involved ir hydrogen-bonding.

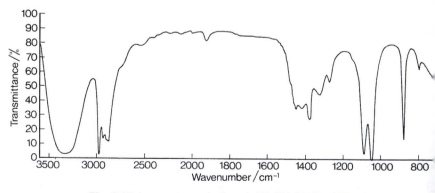

Fig. 3.13 I.r. spectrum of ethanol, CH_3CH_2OH (liquid film)

(iii) N—H *bonds*. Like O—H bonds, N—H bonds show a characteristic absorption at high wavenumber (3500–3300 cm⁻¹). Hydrogen-bonding involving the N—H hydrogen may again cause broadening and a shift towards lower wavenumbers.

iv) C=O *bonds*. Carbonyl-containing compounds show a very characteristic
trong C=O absorption in the range 1800–1600 cm⁻¹, the exact value of
which depends on the structure of the adjacent groups. For simple ketones
nd for aliphatic alkanals the appropriate value is 1725 cm⁻¹ (see Fig. 3.10)
but the wavenumber tends to be slightly higher for simple alkanoyl chlorides,
sters, and anhydrides, and slightly lower than this for amides. The character-
stic wavenumbers are also generally a little lower if the carbonyl group (in
etones, esters etc.) is adjacent to an alkenic double bond (C=C−C=O) or to
. phenyl ring, or if it is involved in hydrogen-bonding (see also Chapters 4
nd 5).

Carbon–oxygen single bonds (C−O) in esters, alkanols, and ethers usually
how a strong absorption in the range 1300–1050 cm⁻¹ [see section (ii)].

Remember that the key absorptions mentioned here and in Table 3.3 are
only some of the characteristic vibrations which may be encountered — the
pectra contain complicating features from other parts of the molecule.

3.4   Examples of i.r.spectra

The following examples illustrate the variation of i.r. spectra with molecular
tructure.

Hexane, CH₃(CH₂)₄CH₃ (Fig. 3.14)

This spectrum reveals very clearly the characteristic aliphatic C−H stretching
modes (just below 3000 cm⁻¹) and bending modes (1500 cm⁻¹) with no indi-
ation of any functional groups.

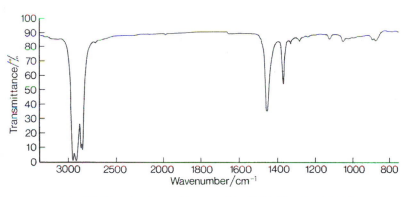

Fig. 3.14 I.r. spectrum of hexane, CH₃(CH₂)₄CH₃ (liquid film)

Pent-1-ene, CH₃(CH₂)₂CH=CH₂ (Fig. 3.15)

The peaks at 900 cm⁻¹ and 3080 cm⁻¹, the alkenic C−H out-of-plane bending
nd stretching modes, respectively, provide conclusive evidence for the
resence of an alkene. Also present are alkyl C−H modes (*ca.* 2900 and
450 cm⁻¹) and the C=C stretching vibration (1640 cm⁻¹). Alkenes without a
erminal double bond show slightly different absorptions below 1000 cm⁻¹:
he details sometimes allow the substitution pattern of the alkene (*cis* or
rans) to be established.

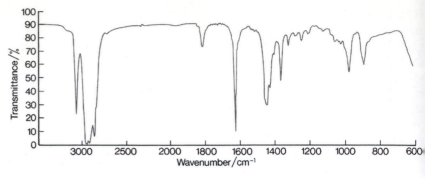

Fig. 3.15 I.r.spectrum of pent-1-ene, $CH_3(CH_2)_2CH{=}CH_2$ (liquid film)

Methylbenzene, $C_6H_5CH_3$ (Fig. 3.16)

In addition to the aromatic and aliphatic $C{-}H$ stretching modes (at 3050 and 2900 cm^{-1}, respectively) there are characteristic peaks from the aliphatic $C{-}H$ bending modes (*ca.* 1500 cm^{-1}) and from out-of-plane aromatic $C{-}H$ bending (*ca.* 700 cm^{-1}). The particular pattern around 700 cm^{-1} is typical of a mono-substituted benzene derivative.

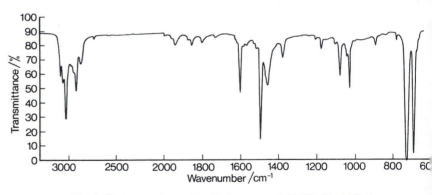

Fig. 3.16 I.r. spectrum of methylbenzene, $C_6H_5CH_3$ (liquid film)

Ethanoic acid (acetic acid), CH_3CO_2H (Fig. 3.17)

This spectrum, from a thin film of the liquid, shows typical $C{-}H$ bending vibrations (*ca.* 1400 cm^{-1}) and the characteristic carbonyl absorption at 1720 cm^{-1}. The broad absorption at 3000 cm^{-1} typifies a hydrogen-bonded OH group.

Ethyl ethanoate (ethyl acetate), $CH_3CO_2C_2H_5$ (Fig. 3.18)

This shows clearly the typical carbonyl stretching absorption at 1740 cm^{-1}; note that this is at slightly higher wavenumber than that observed for ketones. The absorption at 1240 cm^{-1} is a $C{-}O$ stretch, particularly prominent for esters.

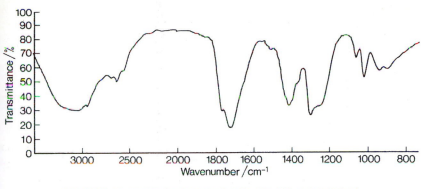

Fig. 3.17 I.r. spectrum of ethanoic acid, CH_3CO_2H (liquid film)

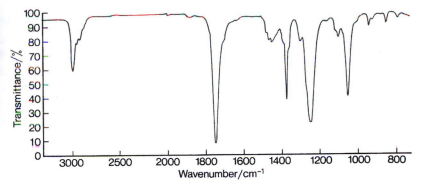

Fig. 3.18 I.r. spectrum of ethyl ethanoate, $CH_3CO_2C_2H_5$ (liquid film)

Diethylamine, $(C_2H_5)_2NH$ (Fig. 3.19)

The broad peak at *ca.* 3300 cm^{-1} in this spectrum is indicative of an N−H group; the broadening suggests that the hydrogen atom takes part in inter-molecular hydrogen-bonding with nitrogen atoms in other molecules. The strong absorption at 1140 cm^{-1} is from a C−N stretching vibration (cf. C−O stretch in alcohols, ethers, etc.).

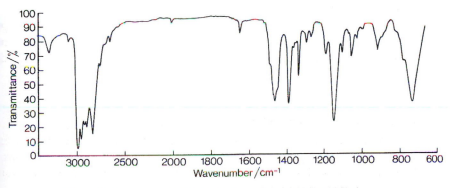

Fig. 3.19 I.r. spectrum of diethylamine, $(C_2H_5)_2NH$ (liquid film)

3.5   Problems

3.1   Calculate the fundamental stretching frequency and wavenumber for a C−H bond, given that the atoms vibrate independently of other groups on the carbon atom and that the force constant is 4.8×10^2 N m^{-1}.

3.2   Calculate the expected separation (in wavenumbers) between the rotational transitions in the far infra-red (or microwave) spectrum of carbon monoxide given that the bond length, r_{CO}, is 0.113 nm.

3.3   Account for the observation that in the i.r. spectrum of deuteriated benzene ($C_6{}^2H_6$), the C−D (C−^{2}H) symmetric stretching mode occurs at *ca.* 2280 cm^{-1} (in undeuteriated benzene the corresponding absorption is at 3050 cm^{-1}).

3.4   Fig. 3.20 is the i.r. spectrum of the compound whose mass spectrum has been given previously (Problem 1.3, page 16) and whose n.m.r. spectrum is described on page 69. Does the i.r. spectrum support your previous assignment?

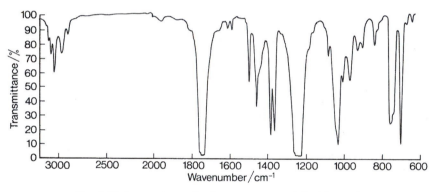

Fig. 3.20 I.r. spectrum of unknown compound: Problem 3.4

3.6   Conclusion

The usefulness of i.r. spectroscopy lies not only in the detailed information which can be obtained for small molecules, but also, for example, in the rapid and fairly inexpensive technique it provides for deciding which functional groups are present in a molecule. It has some advantages over mass spectra (Chapter 1) and n.m.r (discussed in Chapter 5); thus, compared with the former it has the advantage that it can be easily applied to non-volatile samples. Compared with n.m.r. it possesses the ability to give information about all the atoms in a molecule (not just those nuclei with magnetic moments). In practice, all three techniques (with ultra-violet spectroscopy) can usually be used to complement each other in a complete diagnosis.

    The technique also finds increasing applications in a wider context. For example, Fig. 3.21 shows the i.r. spectrum obtained from a thin film of nylon-6,6 (this polymer is prepared by the **copolymerization** reaction of hexane-dioic acid and hexane-1,6-diamine). Infra-red absorptions from C−H

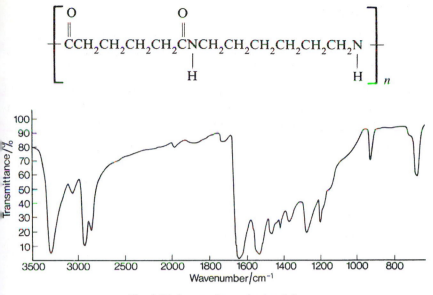

Fig. 3.21 I.r. spectrum of nylon-6,6

stretching (2900 cm⁻¹), amide C=O stretching (1600 cm⁻¹) and N−H stretching (3300 cm⁻¹) help confirm the structure: note that the bonds around the amide function (the **peptide link** CO−NH) are planar; out-of-plane N−H bending accounts for the absorption at *ca.* 700 cm⁻¹.

A thin film of perspex has strong absorptions at *ca.* 2950, 1730, 1450, and 1200 cm⁻¹. What type of polymer structure is consistent with these observations?

A quite different application is provided by a structural investigation of the metal-carbonyl compound $Fe_2(CO)_9$, (Fig. 3.22), which has i.r. absorptions at 2020 and 1830 cm⁻¹. The former absorption is from terminal carbonyl groups (rather similar to carbon monoxide itself, which has $\bar{\nu}$ 2146 cm⁻¹), whereas the latter is from the bridging carbonyl groups (cf. R_2C=O, *ca.* 1700 cm⁻¹).

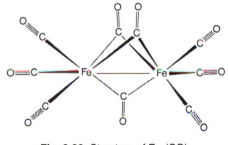

Fig. 3.22 Structure of $Fe_2(CO)_9$

Finally, infra-red spectroscopy may also be employed for the rapid quantitative analysis of the components of a mixture. For example, Fig. 3.23 shows the i.r. spectrum of a sample taken from a car exhaust. Spectra (due to vibrational and rotational transitions) can be attributed to carbon monoxide,

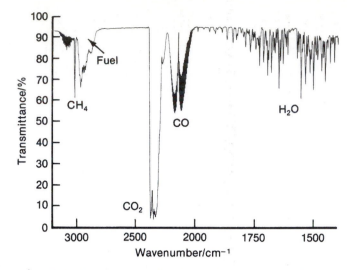

Fig. 3.23 I.r. spectrum of the mixture of gases in a car exhaust

carbon dioxide, water, unspent hydrocarbon fuel, and also methane (formed in the 'cracking' of other hydrocarbons). The amount of any of these in the gaseous emission can be quantified and, for example, continuously monitored during performance tests. A commercial i.r. spectrometer is also available which gives rapid quantitative analysis of different components (e.g. lactose, proteins and fats) in milk samples.

The more rapid acquisition of inra-red spectra can now be achieved using FTIR spectrometers which employ Fourier Transform procedures (see page 71), an approach which also offers advantages in sensitivity.

Further Reading (Chapters 3 and 4)

1.   L. M. Harwood and T. D. W. Claridge, *Introduction to Organic Spectroscopy*, Oxford Chemistry Primers, Oxford University Press, 1997.
2.   D. H. Williams and I. Fleming, *Spectroscopic Methods in Organic Chemistry*, 5th Edition McGraw Hill, 1995.

4 Electronic (ultraviolet-visible) absorption spectroscopy

In addition to the absorption of well-defined amounts of energy to increase its vibrational and rotational energy, a molecule may also absorb energy to increase the energy of its **electrons**. The energy changes involved are considerably greater than those involved for vibration and rotation and correspond to radiation in the *ultraviolet* (u.v.) (λ *ca.* 200–400 nm) and *visible* (λ 400–750 nm) regions of the electromagnetic spectrum.

For radiation of λ 400 nm, $\nu = 7.5 \times 10^{14}$ Hz, the energy of one quantum ($h\nu$) is 5×10^{-19} J. This is the energy which is absorbed by *one molecule* if it absorbs one quantum (or photon) of violet light, and it is obviously a very small quantity. However, for one *mole* of material (in which the number of molecules is the Avogadro constant, L) the total energy absorbed (L quanta) corresponds to $(5 \times 10^{-19}) \times (6 \times 10^{23})$ i.e. *ca.* 300 kJ: this is the same order of magnitude as some *bond enthalpies* (energies) in typical molecules. The interaction of visible or u.v. radiation with molecules can therefore sometimes bring about chemical reactions involving bond breakage: such processes are called **photochemical** reactions.

4.1 Electronic energy changes

The overall energy of a molecule is the sum of contributions from electronic, vibrational and rotational energy, as given by eqn 4.1, where E_{vib} and E_{rot} have exactly the same allowed values as discussed previously (Chapter 3). This illustrated in Fig. 4.1.

$$E = E_{elec} + E_{vib} + E_{rot} \qquad (4.1)$$

In a sample of a stable substance at room temperature, all the molecules will be in the E_0 level (the ground electronic state) but may have different **v** and J quantum numbers. When the appropriate energy is incident it is possible to excite the molecule to the higher energy *electronic* state E_1 (which again has similarly quantized rotational and vibrational energy levels). The excited state arises because of the promotion of electrons to higher energy *molecular orbitals*, exactly as can happen for electrons in atomic orbitals (cf. the sodium flame test, page 21).

This type of absorption explains why some organic and inorganic compounds are coloured — the absorption of energy for the transition $E_0 \rightarrow E_1$ is in the visible region. Other compounds absorb energy in the ultraviolet region and although they do not appear coloured (unless there is also visible absorption) the absorption in this part of the spectrum can be detected. The phenomena of fluorescence and phosphorescence, beyond the scope of this book, are associated with the re-emission of energy (particularly in the visible region) when the molecules in an excited electronic state return to the ground state.

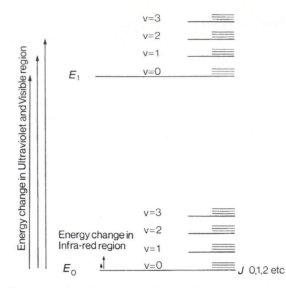

Fig. 4.1 Representation of electronic, vibrational, and rotational energy levels

4.2   Electronic absorption spectroscopy of organic molecules

The experiment

The approach is similar to that employed for infra-red spectroscopy, and many commercial instruments are available. The *source* provides radiation in a continuous range of wavelengths in the u.v. and visible region, as from a heated tungsten filament (e.g. an electric light bulb). Two separate sources are usually required for covering the whole u.v. and visible region. A *prism* or *grating* separates the radiation into component wavelengths and the absorption of the sample at any particular wavelength is measured by the reduction of the signal from a *photoelectric device* when the sample is placed in the incident beam.

Most investigations involve the use of liquid samples, which usually contain the compounds to be studied as solutions in suitable solvents. Modern spectrometers employ a *double-beam* system whereby two identical beams of radiation are generated, one of which passes through the solution under investigation, the other of which passes through an equivalent amount of pure solvent, so that the difference in absorption which is measured is just that due to the molecules of the *solute* in the solution: the *solvent* should be transparent in the region of interest. Tetrachloromethane, hexane, cyclohexane, and ethanol are often employed as solvents, and the solutions are usually contained in quartz glass cells such that the beam of radiation passes through 1 cm of solution (this is known as the path length).

Most spectrometers automatically record *absorption* as a function of *wavelength*. In others you record the absorption at different wavelengths. A colorimeter is a simple type of spectrometer working with a given colour; i.e. at a fixed range of wavelength in the visible region, selected with a suitable filter from a wide range of radiation.

Examples of spectra

Figs. 4.2–4.4 are the electronic absorption spectra of three organic molecules — propanone, benzene, and the indicator methyl red. These are plots of wavelength (λ/nm) against absorbance (A) (sometimes also referred to as optical density). The absorbance is the logarithm of the ratio of the intensity of the incident radiation (I_0) to that of the transmitted radiation (I).

A peak in the spectrum at a given λ corresponds to absorption of energy at this wavelength by the solute molecules; for some molecules, more than one area of absorption is observed, as in Figs 4.2 and 4.3: strong absorptions tailing off into the normally inaccessible region below 200 nm are detected in addition to the higher-wavelength peaks. Propanone and benzene are colourless, since they do not absorb in the visible region, but methyl red absorbs in acid solution in the blue-green region (λ 400–600 nm) and so appears red; in alkaline solution, the absorption is at lower wavelengths and the solution is yellow. The peaks observed are broad in most cases because molecular interactions in the liquid cause the obliteration of the expected vibrational and rotational fine structure, although vibrational fine-structure is clearly apparent in the absorption with λ *ca.* 250 nm from benzene.

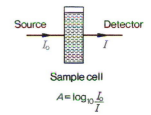

Source   Detector

I_0 I

Sample cell

$$A = \log_{10}\frac{I_0}{I}$$

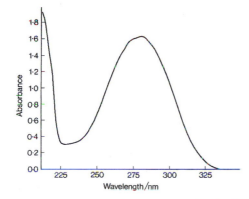

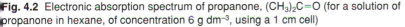

Fig. 4.2 Electronic absorption spectrum of propanone, $(CH_3)_2C{=}O$ (for a solution of propanone in hexane, of concentration 6 g dm^{-3}, using a 1 cm cell)

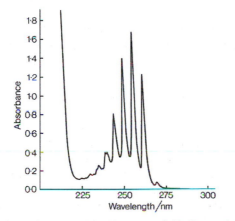

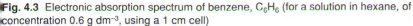

Fig. 4.3 Electronic absorption spectrum of benzene, C_6H_6 (for a solution in hexane, of concentration 0.6 g dm^{-3}, using a 1 cm cell)

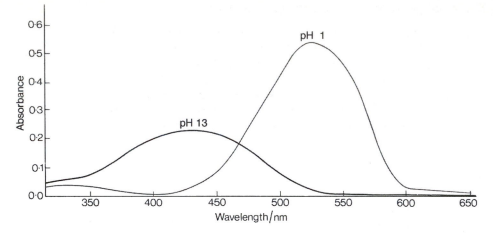

Fig. 4.4 Electronic absorption spectra of aqueous solutions of the indicator methyl red (containing 0.004 g dm^{-3} of the indicator; recorded with a 1 cm cell)

Electronic absorption spectra are usually characterized by two parameters:

(i) the values of the *wavelengths* at which *absorption maxima* occur (λ_{max}): Figs 4.2–4.4 indicate the different values obtained for the different molecules considered; for example, for propanone, the high-wavelength absorption has $\lambda_{max} = 279$ nm; and

(ii) the *extent* of absorption, for a given concentration of compound at any given wavelength (i.e. the *height* of the peak).

As will be seen, the position (λ_{max}) and extent of absorption provide two more characteristic properties of a molecule which depend on its structure.

The Beer–Lambert Law

The extent of absorption at a given wavelength by an absorbing compound in a non-absorbing solvent is found to depend upon the *concentration* of the compound and upon the *path length* of the cell. The **Beer–Lambert Law**, which is generally well obeyed for fairly dilute solutions, expresses the dependence of the absorbance on these two variables (eqn. 4.2). If b is in m and c is in mol m^{-3}, then ε, in m^2 mol^{-1}, is described as the **molar decadic absorptivity** or **molar extinction coefficient**. Thus if b and c are known, and if the experiment gives a value for A (the spectrometer records this), ε can be calculated. This is usually quoted for the wavelength of maximum absorption (i.e. at λ_{max}).

Once ε has been determined, the value of A for a given solution of a known compound can be used to determine c, the concentration of that compound in the solution. This behaviour, the basis of the Beer–Lambert Law, means that u.v.-visible spectrophotometry provides an excellent method for quantitative as well as qualitative analysis.

$$A = \log_{10} \frac{I_0}{I} = \varepsilon c b \qquad (4.2)$$

where A is the absorbance
b is the path length
c is the concentration
ε is a constant for a particular compound at chosen wavelength

4.3 The relationship of λ_{max} and ε_{max} to structure

The electronic absorption spectra of a variety of organic compounds show that only certain types of molecule exhibit absorption in the u.v.-visible range (λ 200–750 nm). These must contain double or triple bonds (or, in some cases, lone-pairs of electrons), which are essentially responsible for the absorption; these fragments are called **chromophores**. When two or more chromophores are adjacent to each other (the groups are then said to be **conjugated**) the absorptions become more pronounced (higher ε_{max}) and occur at lower energy (greater λ_{max}).

This behaviour can be understood in terms of the types of molecular orbitals involved in the electronic excitations. There are three types of molecular orbital, and the electrons in these orbitals have somewhat different environments. First, there are the electrons in the σ-orbitals which constitute the bonding framework of a molecule (e.g. the electrons in the 4 C−H bonds in methane, CH_4). These orbitals are formed from overlap of s, sp, sp^2, and sp^3 orbitals. Second, there are electrons in π-orbitals, formed from laterally overlapping atomic p-orbitals in compounds such as benzene and ethene (ethylene). Third, there are *lone-pair* electrons in orbitals on atoms like oxygen, nitrogen, etc.; these are called non-bonding or n-electrons. The carbonyl group (in a ketone, say) contains all three types of molecular orbital.

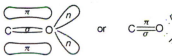

When two electrons in atomic orbitals are brought together to form a bond (i.e. when a filled molecular orbital is produced) a higher energy *anti-bonding* orbital is also formed which is empty in the ground state of the molecule (cf. empty high-energy atomic orbitals). When excitation takes place, an electron from one of the filled orbitals (σ, π or n) becomes excited to a vacant anti-bonding orbital (σ^*, π^*) so that a new *excited* state is reached. Since various excitations are possible, there are various possible absorptions, corresponding to the transitions $n \rightarrow \sigma^*$, $\sigma \rightarrow \pi^*$, etc.

The *approximate* relative energies of typical σ, π, n and anti-bonding orbitals are indicated in Fig. 4.5; as expected, the σ-electrons are the most tightly bound (most energy is needed to excite them). The order of decreasing energy for the absorptions is as follows:

$$\sigma \rightarrow \sigma^* > \sigma \rightarrow \pi^* \sim \pi \rightarrow \sigma^* > \pi \rightarrow \pi^* \sim n \rightarrow \sigma^* > n \rightarrow \pi^*$$

Of all these possible transitions, only those of the last three types normally account for absorption in the u.v.-visible region, the others requiring too great an energy. This then explains why only molecules with n or π electrons give rise to characteristic u.v. and visible spectra whereas alkanes, for example, show no absorption in this region.

Ethanol absorbs radiation of wavelength *ca.* 200 nm and below. This absorption derives from the $n \rightarrow \sigma^*$ absorption by the lone-pair electrons on the oxygen atom. Ethanol is transparent above this wavelength and finds use as a solvent for u.v.-visible studies on molecules with higher-wavelength absorptions. For a variety of organic molecules which contain chromophores (double bonds etc.) and which also give absorption in the u.v.-visible region, Table 4.1 indicates the measured values of ε_{max} and λ_{max}, together with the types of transition.

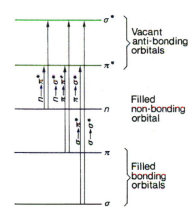

Fig. 4.5 Approximate relative energies for electrons in different types of molecular orbital in organic compounds (not to scale)

Table 4.1 Characteristic u.v.–visible absorptions for organic compounds

(a) **Molecules with single chromophores**		Solvent	λ_{max}/nm	Transition type	ε_{max}/m^2 mol$^-$
Propanone	$(CH_3)_2CO$	hexane	188	$\pi \rightarrow \pi^*$	90.0
			279	$n \rightarrow \pi^*$	1.5
Ethyl ethanoate (ethyl acetate)	$CH_3C(O)OCH_2CH_3$	water	204	$n \rightarrow \pi^*$	6.0
Pent-1-ene	$CH_3CH_2CH_2CH{=}CH_2$	hexane	190	$\pi \rightarrow \pi^*$	1000
Nitromethane	CH_3NO_2	hexane	278	$n \rightarrow \pi^*$	1.7
(b) **Conjugated molecules**					
Buta-1,3-diene	$CH_2{=}CH{-}CH{=}CH_2$	hexane	217	$\pi \rightarrow \pi^*$	2100
Butenone	$CH_2{=}CH{-}C(O)CH_3$	ethanol	219	$\pi \rightarrow \pi^*$	360
			324	$n \rightarrow \pi^*$	2.4
Benzene	C_6H_6	hexane	184	all $\pi \rightarrow \pi^*$	6000
			203		740
			255		20
Acetophenone	$C_6H_5C(O)CH_3$	ethanol	199	all $\pi \rightarrow \pi^*$	2000
			246		1260
			279		100
			320		4.5
Nitrobenzene	$C_6H_5NO_2$	hexane	252	all $\pi \rightarrow \pi^*$	1000
			280		100
			330		12.5

For propanone (see Fig. 4.2) the peak at $\lambda = 188$ nm (for which $\varepsilon = 90$ m mol^{-1}) is responsible for absorption at the low-wavelength end of the observed spectrum, and the peak at $\lambda = 279$ nm ($\varepsilon = 1.5$ m^2 mol^{-1}) is also clearly visible. These are due to transitions involving the electrons in the carbonyl-group double bond ($\pi \rightarrow \pi^*$) and the oxygen's lone-pair electrons ($n \rightarrow \pi^*$), respectively. The characteristic values are slightly sensitive to the solvent used.

These absorptions do not depend to any marked extent on the nature of the alkyl groups attached to the carbonyl function; thus for a variety of ketones and alkanals, the u.v. spectra recorded for hexane solutions show the following values of λ and ε for the $n \rightarrow \pi^*$ absorption: butanone (279 nm, 1.6 m^2 mol^{-1}) cyclohexanone (285 nm, 1.4 m^2 mol^{-1}), ethanal (acetaldehyde, 293 nm 1.2 m^2 mol^{-1}), propanal (290 nm, 1.8 m^2 mol^{-1}). For carbonyl-containing compounds of different chemical type (e.g. alkanoate esters), the absorptions are, however, characteristically different (see Table 4.1).

When a carbon–carbon double bond is present, a characteristic $\pi \rightarrow \pi$ absorption is observed (Table 4.1 gives data for pent-1-ene): what is particularly noticeable here is the larger value of ε_{max} (which means that a more dilute solution of the compound is needed, compared with the ketones, to obtain the same amount of absorption). This type of increase often occurs when a transition takes place between states of similar type (e.g. $\pi \rightarrow \pi$ compared with $n \rightarrow \pi^*$).

When two chromophores in a molecule are adjacent (conjugated) it is generally found that the energy needed for absorption *decreases* (i.e. λ_{max}

increases) and the extent of absorption (ε_{max}) *increases* compared with the values for separate groupings. This is illustrated for the diene and the unsaturated ketone listed in Table 4.1. The effect is particularly marked for benzene (and other aromatic compounds) which have extended π-systems. These molecules can also be distinguished since there are several $\pi \to \pi^*$ absorptions (owing to the existence of several π and π^* orbitals).

When a benzene ring and another chromophore are conjugated then the characteristic absorptions are shifted to even longer wavelengths: an example is provided by Fig. 4.6, which shows the electronic absorption spectrum of acetophenone, $C_6H_5COCH_3$ (there is also a weak absorption at higher wavelength; the details are given in Table 4.1). Substitution of alkyl groups on aromatic and alkenic chromophores leads to small increases in λ_{max}.

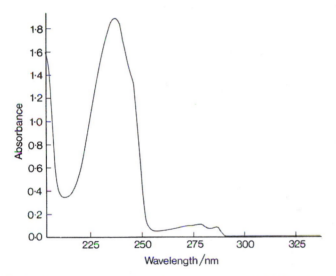

Fig. 4.6 Electronic absorption spectrum of acetophenone, $C_6H_5COCH_3$ (for a solution in hexane using a 1 cm cell; see Problem 4.3)

The increased wavelength of absorption for *conjugated* molecules (i.e. when chromophores are adjacent to each other) can lead, if enough chromophores are present, to an absorption in the visible region. This occurs, for example, for aromatic compounds containing fused rings, for some 1,2-diketones (which are yellow), for large molecules (with delocalized π-electrons) used as pH indicators (the conjugation and hence the colour depends upon the ionization of groups in the molecule) and for molecules containing chains of double bonds.

For example, β-carotene (which occurs in carrots) is orange and has $\lambda_{max} = 450$ nm, $\varepsilon_{max} = 15,000$ m^2 mol^{-1}. When ε_{max} is very high, only very dilute solutions are needed for detection of the absorbing molecules, and the technique becomes a very sensitive method indeed for the detection of absorbing species (for example, absorption could be detected for a solution made up from less than 0.01 mg of β-carotene).

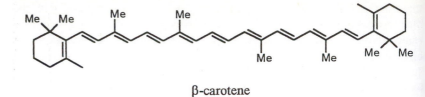

β-carotene

4.4   Some applications of u.v. and visible absorption spectroscopy

Structural analysis

One important use of electronic absorption spectroscopy is the recognition o
chromophores or groups of chromophores in organic molecules, by the meas
urement of λ_{max} and ε_{max} for the various absorption peaks. This information
usually allows the type of molecule to be determined and, particularly whe
used in conjunction with other spectra, may provide valuable assistance wit
the determination of the exact molecular structure.

This branch of spectroscopy has been particularly useful in the structura
investigation of steroids, which are biologically important molecules. A
example is the hormone testosterone. The absorption at 241 nm ($\varepsilon = 1600$ m
mol^{-1}) is characteristic of adjacent C=C and C=O double bonds in this typ
of cyclic structure; it has proved possible in examples like this to use the u
data to give an indication of the detailed structure around the chromophores.

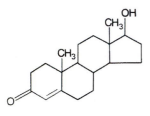

Quantitative analysis

We can often use the Beer–Lambert Law to relate the absorbance from a par
ticular substance to its concentration (if ε is known), with many application
in quantitative analysis. The advantages of this method are that very low con
centrations can be reliably obtained and that the *rate* of change of the absorp
tion at a given wavelength can easily be monitored (e.g. if the compound i
involved in a chemical reaction). Some typical applications are as follows.

(i) *Kinetic Investigations.* Most spectrometers produce a plot of absorbanc
against λ but also allow the absorbance at one particular wavelength to b
plotted as a function of time. This facility leads to a plot of concentratio
against time for either the loss of a reactant or the build-up of a product, a
long as the absorption is characteristic of only the single component unde
investigation. Thus, the spectrometer not only provides vital informatio
about the nature of the product (or products) of a reaction, it may also allo
the rate of the reaction to be followed. Both pieces of information can aid th
elucidation of the reaction mechanism concerned.

(ii) *Keto-enol tautomerism.* Both the u.v. spectrum (Fig. 4.7) and the n.m.r
spectrum of pentane-2,4-dione (acetylacetone, $CH_3COCH_2COCH_3$) indicate
that this molecule does not exist simply as the molecular formula suggests —
i.e with two keto groups. The n.m.r. spectrum provides detailed evidence fo
two different forms of the molecule present in equilibrium: these are the **ket**
and **enol tautomers** of the molecule (the phenomenon is called **tautomerism**)

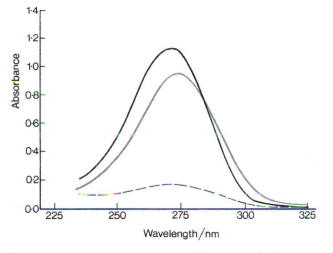

Fig. 4.7 Electronic absorption spectra of solutions (10^{-4} mol dm^{-3}) of pentane-2,4-dione, $CH_3COCH_2COCH_3$, in hexane (—), ethanol (—) and water (---); the cell has a path length of 1 cm (see Problem 4.4)

Figure 4.7 shows the absorption for 10^{-4} mol dm^{-3} solutions of acetyl-acetone in hexane, ethanol, and water. First, the extent of absorption suggests that a simple carbonyl-type structure is not responsible for the peak [for example, a much greater concentration of propanone is necessary to obtain a strong ($n \rightarrow \pi^*$) absorption: see Fig. 4.2]. A $\pi \rightarrow \pi^*$ transition associated with a *conjugated* structure is more likely. Second, the variation in the extent of absorption with solvent is surprising (it certainly does not occur for a simple ketone).

The explanation is that the absorption at λ = 270 nm is due to the *enol* form (cf. absorption for compounds of the type C=C–C=O, Table 4.1) and it can be demonstrated (e.g. by n.m.r. spectroscopy) that for a solution in ethanol there is approximately 73% of the enol and 27% of the keto form together in equilibrium (the percentage of the former accounts for the absorption with λ_{max} = 270 nm, absorbance = 0.96 from the 10^{-4} mol dm^{-3} solution of the diketone in ethanol). The absorptions for aqueous and hexane solutions imply that in water there must be a much smaller proportion of enol than when ethanol is the solvent and that in hexane solution there is correspondingly more enol. In hexane the internally hydrogen-bonded enol form is preferred (the dotted line in the structure indicates a hydrogen bond) whereas in aqueous and alcoholic solutions the formation of *intermolecular* hydrogen bonds between the carbonyl group and the water molecules (or alcohol molecules) stabilizes the keto form.

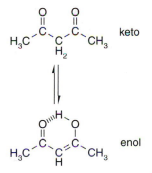

Indicators

Fig. 4.8 shows the variation in the absorption spectrum of a dilute aqueous solution of the indicator methyl red as the pH is altered. The indicator is red in acid (λ_{max} = 520 nm, see Figs 4.4 and 4.8) and yellow in alkali (λ_{max} = 425 nm), these being the colours of the acid and base forms.

At pH 1, the indicator is essentially all in the acid (HA) form; at pH 13, it is essentially all in the base (A$^-$) form. For intermediate pH values, both HA

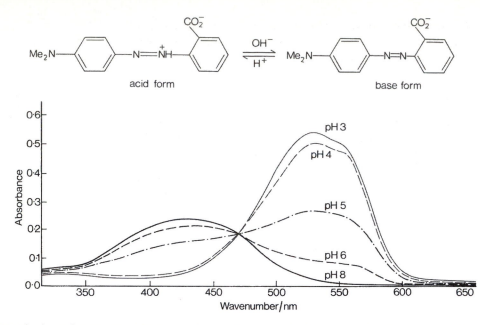

Fig. 4.8 Electronic absorption spectra of aqueous solutions of methyl red (all containing 0.004 g dm⁻³ of the indicator) at various pH values

and A⁻ are present (to an extent which is governed by the pH and by the dissociation constant of the indicator, K_a). The heights of the 'acid' peak ($\lambda = 520$ nm) and the 'base' peak ($\lambda = 425$ nm) can be used, together with the 100% 'acid' and 'base' absorptions, to determine the concentrations of HA and A⁻ at any given pH. This then leads to a measurement of K_a. Alternatively, if K_a is known, then measurement of [HA] and [A⁻] gives a value for the pH. As will be appreciated, these measurements serve to quantify the procedure whereby, in a titration, the eye responds to the change in colour of a solution when a predominance of HA, say, is changed to a predominance of A⁻.

Metal ions and complexes

The application of u.v.-visible absorption spectroscopy also allows the determination of absorption maxima and molar decadic absorptivities for inorganic ions with electronic absorptions in this part of the spectrum [e.g. for the absorptions which account for the purple colour of the manganate (VII) ion (permanganate, MnO_4^-) and for the yellow colour of the dichromate (VI) ion ($Cr_2O_7^{2-}$)]. This type of detailed information can then be useful in several contexts. For example, the ions MnO_4^- and $Cr_2O_7^{2-}$ could be identified as present in a solution from the appearance of their characteristic absorption spectra. Further, the inorganic anions mentioned above could be formed by the oxidation of solutions containing trace quantities of, say, Mn(II) and Cr(III), and the extent of absorption (the absorbance) at the appropriate wavelengths could be use to calculate the concentrations of the ions present (this can often be done for mixtures of ions from a single spectrum). In addition, there is considerable interest in the detailed analysis of absorptions from metal ions and complexes in terms of the electronic structure (and hence the possible electronic transitions) of the molecules or ions concerned. Particularly useful information can be obtained about transitions and energy levels involving *d*-electrons in transition metal ions.

4.5   Problems

4.1   Fig. 4.2 is the u.v. spectrum of a solution of propanone in hexane. Calculate the molar decadic absorptivity (in m^2 mol^{-1}) for the peak with $\lambda = 279$ nm and compare your result with that on page 48.

4.2   Fig. 4.3 was recorded for a solution of benzene in hexane. Calculate the molar decadic absorptivity (m^2 mol^{-1}) for the peak with $\lambda = 255$ nm.

4.3   For a solution of acetophenone ($C_6H_5COCH_3$) in hexane the molar decadic absorptivity of the peak with $\lambda_{max} = 279$ nm is 100 m^2 mol^{-1}. Calculate the concentration of the solution whose u.v. spectrum is shown in Fig. 4.6.

4.4   (a) Calculate the apparent molar decadic absorptivity for λ_{max} for pentane-2,4-dione in ethanol from Fig. 4.7, which shows the spectrum for a solution containing 10^{-2} g dm^{-3} of the di-ketone.

(b) The proportion of the enol of pentane-2,4-dione in solution in ethanol has been estimated as 73%. From Fig. 4.7 calculate the proportion of enol for solutions in (i) hexane, and (ii) water.

4.5   Explain why aniline (phenylamine, $C_6H_5NH_2$) shows absorption maxima at approximately 230 and 280 nm ($\varepsilon = 860$ and 143 m^2 mol^{-1}, respectively), whereas salts of the anilinium (phenyl-ammonium) cation ($C_6H_5NH_3^+$) have absorptions at *ca.* 200 and 250 nm ($\varepsilon = 750$ and 16 m^2 mol^{-1}, respectively).

4.6   From Fig. 4.8, estimate the dissociation constant K_a of methyl red.

5 Nuclear magnetic resonance spectroscopy

Since the first nuclear magnetic resonance (n.m.r.) experiment was successfully demonstrated in 1946, the technique has become so powerful that it plays an essential role in modern research. Indeed, n.m.r. and mass spectroscopy now have no real rivals as the techniques most widely applicable for the solution of structural problems, and their use is now a routine matter. In the form of MRI (magnetic resonance imaging), n.m.r. has moved from the realm of research to become an essential feature of modern medicine.

5.1 The n.m.r. experiment

At the heart of each n.m.r. experiment lies the nucleus of an atom. An atom can be thought of as a sea of negatively charged electrons surrounding a positively charged nucleus which is itself composed of protons (positively charged) and neutrons. It has been found that nuclei which contain an odd number of protons or an odd number of neutrons (or both) possess an extra property, in addition to their charge, which can be demonstrated by a suitable experiment. They can be shown to have a **magnetic moment**, which means that in the simplest case they behave like tiny bar magnets. This phenomenon can be demonstrated by detecting the energy of interaction when they are placed between the pole-pieces of a magnet. As happens when a simple bar magnet is placed in the magnetic field of a second magnet, the distinction between attractive and repulsive interactions can be detected.

The hydrogen atom, ^{1}H, has a single proton for its nucleus and hence has a magnetic moment. In the presence of an externally applied magnetic field the magnetic moment experiences an interaction which results in it becoming aligned either parallel to or opposed to the direction of the applied field. This picture of two allowed orientations is also valid for ^{13}C, ^{19}F, ^{31}P, and ^{15}N, but for nuclei like ^{2}H the situation is more complex.

The two possible alignments of the magnetic moment of the hydrogen nucleus ^{1}H are represented diagrammatically in Fig. 5.l. These arrangements have different *energies,* and an exact amount of energy is necessary to twist the magnet from one position to the other, i.e. from the attractive to the repulsive situation.

Now for a magnetic moment μ the component in the direction of the applied field is $+\mu$ (designated α) or $-\mu$ (designated β), depending on whether it is parallel or anti-parallel to it. The energies of these two arrangements of the magnet are $-\mu B_0$ (aligned) and $+\mu B_0$ (opposed) respectively, where B_0 is the magnetic flux density of the applied field. The energy difference between the two arrangements is $2\mu B_0$ and this amount of energy (as radiation of the corresponding frequency, see page 19) is necessary to invert

It will be helpful to remember that a magnetic moment (or field) is associated with a body which is charged and in motion (cf. the magnetic field from electrons moving in a wire). This requires the ^{1}H nucleus to be spinning in one of two directions.

Fig. 5.1 The two allowed alignments of the nuclear magnetic moment of a hydrogen atom in an applied magnetic field

the proton's magnetic moment from the position of lower energy to that of higher energy. The exact condition is then:

$$\Delta E = h\nu = 2\mu B_0 \qquad (5.1)$$

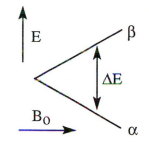

The energy difference between an α spin state and a β spin state depends on the size of the external magnetic field B_0

This equation relates μ, the magnetic moment of the proton, the magnetic flux density B_0 (i.e. the applied field strength), and the frequency ν which has to be employed before energy can be absorbed by the proton. The value of ν necessary to satisfy equation 5.1 clearly depends on the magnitude of the applied field. For values of B_0 typical of n.m.r. experiments, ν is in the **radio-frequency** region of the electromagnetic spectrum. This means that the values of ΔE appropriate to n.m.r. are much smaller than the energy differences associated with electronic, vibrational, and rotational changes.

The simplest way in which the n.m.r. experiment may be carried out is to vary the magnetic field and keep the applied frequency ν constant in a search for the exact condition when absorption of energy leads to 'flipping' of the hydrogen atom's magnetic moment. In this experiment a radiofrequency oscillator is the source of electromagnetic radiation of fixed frequency ν and an electromagnet is employed to generate the variable magnetic field (see Fig. 5.2). The external magnetic field strength B_0 is then varied until the radiofrequency radiation matches the energy difference ΔE and it is absorbed (at resonance); at this point an imbalance is produced in a radiofrequency bridge, and the resulting signal can be amplified and fed to a recorder as a plot of absorption of energy against frequency.

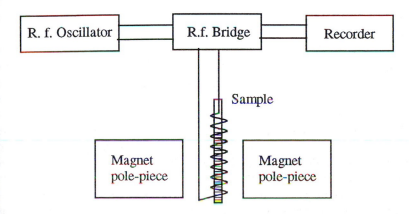

Fig. 5.2 Basic features of a continuous-wave n.m.r. spectrometer

This approach suffers from the limitation that it takes several minutes to scan the magnetic field range and it has now largely been overtaken by the Fourier transform method. This method of data measurement offers many benefits and is discussed later in Section 5.6.

In practice, signals can be readily detected for the hydrogen atoms in a small quantity (*ca.* 0.5 cm^3) of a suitable solvent which would normally have no hydrogen atoms and therefore no absorption in this region. Signals can be detected from substrates at low concentration (*c. sub*-millimolar) which means that n.m.r. spectra can be recorded even when only a few milligrams of compound are available.

In a typical hydrogen-containing sample there will be a large number of ^{1}H nuclei, distributed between the two previously described energy levels; the ratio of the number in the upper energy level (N_u or N_β) to the number in the lower energy level (N_l or N_α) is given by the Boltzmann Distribution (eqn 5.2). Here, T is the Kelvin temperature and k is Boltzmann's constant.

$$\frac{N_u}{N_l} = \frac{N_\beta}{N_\alpha} = e^{-\Delta E/kT} = e^{-2\mu B_0/kT} \quad (5.2)$$

This relationship is generally applicable to the statistical distribution of particles between possible energy levels, and it tells us, for example, that for an energy difference (ΔE) which is large compared with the typical thermal energy of the molecules ($\sim kT$), the lower energy level is much more highly populated. However, for hydrogen atoms at room temperature in a typical magnetic field, ΔE is considerably smaller than kT: the ratio N_u/N_l is nearly unity and the population of hydrogen atoms is almost equally divided between the two energy levels. For a million hydrogen atoms, there will be just a few more in the lower level than in the upper level. When irradiation of the sample with radiation at the resonance frequency (ν) takes place, transitions in both upward and downward directions occur (that is, absorption and emission take place). Overall **absorption** results because of the slight excess of nuclei in the lower level.

The smaller the ratio N_u/N_l, the greater is the sensitivity (i.e. the greater the excess in the lower level); as will be seen from eqn 5.2, this means that a higher applied field (and higher ν, therefore) provides an advantage in sensitivity.

5.2 ^{1}H n.m.r. spectra of organic molecules

On the basis of the previous discussion it would be anticipated that, since n.m.r. is a nuclear phenomenon, the resonance condition for different hydrogen atoms in a variety of molecules should not be affected by the electronic environment of each. To a certain extent this appears to be true; that is, for a fixed frequency (ν), hydrogen atoms in a variety of molecules absorb energy at approximately the same value of B_0. However, minor differences of crucial importance do exist.

For example, when a low-resolution n.m.r. spectrum of *ethanol*, CH_3CH_2OH, is recorded it is found that three distinct resonances occur at slightly different field strengths (the field differences between the resonances are much smaller than the magnitude of the field). The n.m.r. spectrum of ethanol is shown in Fig. 5.3. The areas under the peaks are in the ratio 1:2:3, and we can conclude that the OH hydrogen atom, the two CH_2 hydrogen atoms, and the three CH_3 hydrogen atoms have separate resonances.

Similar observations are made for other organic molecules. For example, the spectrum of *2-methylpropan-2-ol* (t-butyl alcohol, $(CH_3)_3COH$), shows

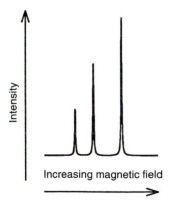

Fig. 5.3 Low resolution ^{1}H n.m.r. spectrum of ethanol CH_3CH_2OH

two peaks in the ratio 9:1, and *diethyl ether*, $CH_3CH_2OCH_2CH_3$, exhibits a spectrum with two peaks in the ratio 3:2. In this case, the two methyl groups are in equivalent environments, but this is different from the environment of the two methene (CH_2) groups.

Fig. 5.4 shows the two-peak spectrum from *ethanal* (acetaldehyde, CH_3CHO), and illustrates the integration trace (upper curve). The height of each step in this trace is proportional to the area under the appropriate resonance and hence to the number of hydrogen atoms in each group. For this example, the ratio is 1:3, corresponding to the separate resonances of CHO and CH_3 hydrogen atoms, respectively.

Hydrogen atoms which undergo absorption of energy at different fields are said to have different **chemical shifts**, which depend on the environments of particular atoms in a molecule.

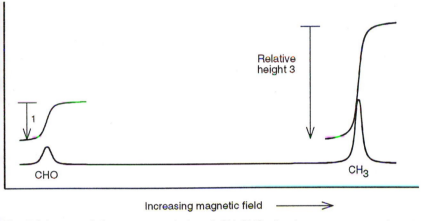

Relative height 3

CHO

CH3

Increasing magnetic field

Fig. 5.4 Low-resolution spectrum of ethanal, CH_3CHO, showing resonances and integration trace

Chemical shifts

The observation that in CH_3CH_2OH, for example, the three types of hydrogen atom absorb at slightly different applied magnetic fields suggests that somehow each type of hydrogen nucleus does not experience exactly the same magnetic field B_0. In practice, it can be shown that in the presence of the applied field small *local* magnetic fields are induced in the neighbourhood of the nuclei. Each nucleus now experiences an effective field,

$$B_{effective} = B_0 + B_{local}$$

and, since B_{local} is found to be proportional to B_0, we can write

$$B_{effective} = B_0(1 + \sigma)$$

Clearly, the size and direction of σ (a measure of the chemical shift) determines the applied field that is needed to achieve the unique resonance condition ($h\nu = 2\mu B_{effective}$) for each type of hydrogen.

It is known that the local fields arise from electron circulations induced by the applied field, and two rather different cases can be distinguished.

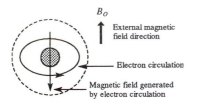

Fig. 5.5 The production of local fields by induced electron circulation

(i) Fig. 5.5 indicates the direction of electron circulation round a nucleus that is induced by the external magnetic field for a spherical electron cloud (this corresponds effectively to an electron in the $1s$-orbital on the hydrogen atom). The electron motion creates an induced magnetic field which opposes the main field. This results in the nucleus experiencing a smaller overall field than that applied (the nucleus is said to be *shielded* from the main field) and the resonance condition can now only be achieved with a higher applied field.

For CH_3CH_2OH the extent of shielding differs for each type of hydrogen atom owing to different *electron densities* around the CH_3, CH_2, and OH hydrogen atoms. For the OH hydrogen atom the electron density around the nucleus will be relatively low owing to the adjacent electronegative oxygen atom. However, for the CH_2 and CH_3 hydrogen atoms, which are progressively further from the oxygen atom, the electron density will be progressively greater. The CH_3 hydrogen atoms, being most highly shielded, are in resonance at the highest field (Fig. 5.3).

In general, we can relate the magnetic fields necessary for resonance of different hydrogen atoms to the structure of a molecule and, in particular, to the electron-withdrawing properties (negative inductive effect) of the atoms present.

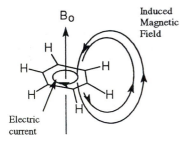

Fig 5.6 Induced electron circulation in benzene: the ring-current effect

(ii) The second type of electron circulation which contributes to local fields in appropriate cases is that which can be induced in molecules containing double bonds (i.e. molecules containing electrons in π-orbitals). A particularly clear example is provided by the marked effect when delocalization around an aromatic ring is possible. This is called a ring current and is illustrated for benzene in Fig. 5.6.

When the benzene molecule is oriented perpendicularly to the applied field, electron circulation is induced in the overlapping p(π)-orbitals and this leads to the production of local magnetic fields around the molecule. These local fields augment the applied field at the hydrogen atoms, so that they resonate at a lower external field strength. The hydrogens are said to be *deshielded*. The rapid motion of the molecules means that only for a fraction of the time is the molecule in the particular orientation depicted; nevertheless the ring-current contribution is a dominant effect and is a readily applied criterion used to distinguish aromatic protons.

This type of effect also contributes to the observed chemical shifts for hydrogen atoms close to carbonyl, alkynic, and certain other groups.

The measurement of chemical shifts

Measurements of chemical shifts are not quoted in field units, because, as we have seen, the field difference (ΔB) between two peaks depends upon the applied field (B_0) of the spectrometer employed (see page 57).

The following procedure provides a way round this problem and leads to an acceptable universal scale. First, a suitable reference compound is chosen; tetramethylsilane (TMS), $Si(CH_3)_4$, is widely used for ^1H spectra as it is inert and has a single resonance at a position which does not often overlap with other signals (there is only one type of hydrogen atom in the molecule). Then, the n.m.r. spectrum of the compound under investigation, with TMS added, is recorded. There will be a field difference ΔB between the absorption of a hydrogen nucleus in the sample and the reference; $\Delta B/B_0$ gives a measure of the chemical shift which is *independent of the applied field*. The resulting numbers are very small indeed and it is more convenient to multiply them by 10^6 and to refer to the resulting measure of chemical shift, δ, in parts per million (p.p.m.).

$$\delta \, / \, p.p.m. = \frac{\Delta B}{B_0} \times 10^6$$

This leads to a scale of chemical shifts with $\delta = 0$ for the TMS hydrogen atoms and with most other hydrogen atoms in organic molecules in the range δ 10–0 (Fig. 5.7). For example, the chemical shift of the hydrogen atoms in benzene is 7.25, and the shifts of hydroxyl, methene, and methyl hydrogen atoms in ethanol are δ 5.20, 3.65, and 1.20 respectively (see Figure 5.3).

Chemical shifts, expressed as δ-values, are characteristic of the hydrogen nuclei in the compounds concerned and are independent of the spectrometer. Further, it is quite rare to find a marked dependence of the δ-value on solvent. Normally chemical shifts are also independent of temperature, but sometimes chemical exchange processes interfere and more complex effects are observed.

Relationship between chemical shifts and molecular structure

The dependence of δ upon molecular structure is understandable in terms of the magnitudes of the local fields produced by the electron circulation effects described earlier. For example, although a CH_3 group in an alkane has a characteristic δ value of approximately 0.9, substitution in the molecule of a group with a negative inductive effect lowers the electron density round the hydrogen atom and hence increases the δ-value.

This increase is especially marked when the substituent is highly electronegative. Thus, along the series of halogeno-alkanes CH_3X (X = I, Br, Cl, and F) the chemical shift of the methyl group hydrogen is progressively increased (see Table 5.1): the highest δ value for the hydrogen atoms in fluoromethane indicates that the hydrogen atoms in this molecule are the least shielded from the external field (as expected, because fluorine is the most electron-attracting substituent).

The effect becomes even more pronounced when more than one electronegative element is present; for the hydrogen atom in $CHCl_3$, for example, $\delta = 7.29$.

Since the chemical shift of a hydrogen atom depends on its immediate environment within a molecule, the resonance positions of hydrogen atoms in compounds of different structural types prove diagnostically useful; some approximate values for typical groups are given in Table 5.2.

The increase in the δ-values along the series CH_3-C, CH_3-N, and CH_3-O, [in alkanes, amines, and methoxy compounds (methyl ethers), respectively] again reflects the increasing negative inductive effect of the substituent group.

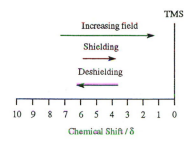

Fig. 5.7 The δ-scale of chemical shifts

Table 5.1 Chemical shifts/δ of hydrogen atoms in a series of halogenomethanes

CH_3F	4.25
CH_3Cl	3.05
CH_3Br	2.70
CH_3I	2.25

Table 5.2 Chemical shifts/δ of hydrogen atoms in organic compounds

Group	Type of compound	Chemical shifts/δ*
H_3C-C	alkane	0.9
$C-CH_2-C$	alkane	1.3
$H_2C-C=C$	alkene	1.6
$H_3C-\overset{\overset{O}{\|}}{C}-O-$	ester, acid	2.0
$H_3C-\overset{\overset{O}{\|}}{C}$	ketone	2.1
H_3C-N	amine	2.3
H_3C-O	methyl ether	3.3
$H_3C-O-C\overset{O}{}$	methyl ester	3.7
$CH_2=C$	alkene	4.7#
$H-C$	arene	7.3#
$\overset{\overset{O}{\|}}{C}-H$	alkanal	9.7#

* Typically ± 0.1; substituent effects may be greater, especially for compounds indicated #

Methyl esters of alkanoic acids, RCO_2CH_3, as might be expected, resemble methoxy compounds (methyl ethers) and have δ (CH_3) in the range 3.6–3.8. In contrast, both methyl ketones and alkyl esters of ethanoic acid (acetic acid), CH_3CO_2R, generally show a methyl group absorption at 2.0, this being the characteristic chemical shift of a methyl group adjacent to a carbonyl group.

Groups which have an inductive effect but which are further removed in the molecule have a decreased but sometimes noticeable effect: an example is the δ-value of 1.20 for the methyl-group hydrogen atoms in ethanol (raised from *ca.* 0.9 by the effect of oxygen, but not as high as the δ-value of *ca.* 3.3 for molecules in which a CH_3 group is directly attached to the oxygen atom).

Methene hydrogen atoms (CH_2) have slightly higher δ-values than similarly placed CH_3 hydrogen atoms; the characteristic δ-value for a $-CH_2-$ group in an alkyl chain is 1.3.

A hydrogen atom attached to a carbon atom in the double bond of an alkene has a higher δ-value (typically about 4.7) than hydrogen atoms in saturated analogues. This shift is associated with the sp^2 rather than the sp^3 hybridization of the alkenic carbon to which the hydrogen atom is bonded. Since few other types of hydrogen atom absorb in this area of the n.m.r. spectrum, absorptions at δ *ca.* 5 have particular diagnostic value.

For benzene and other aromatic compounds the ring-current effect (Fig. 5.6) explains the unusually high δ-values observed.

The lower δ-value for a hydrogen atom attached to an alkyne triple bond (*ca.* 2.0) also results from electron circulation (in the triple bond, around the molecular axis) which in this molecule provides *a shielding* effect at the hydrogen atoms. You may like to work out why the effect is opposite in direction to that observed for the hydrogen atoms in benzene.

When an unknown compound is studied with the aid of n.m.r. spectroscopy, the positions of the absorptions (i.e. the δ values) give a fairly clear indication of the local environment of each type hydrogen atom in the molecule, and the data given in Table 5.2 therefore prove diagnostically useful. In addition, remember that the integration trace gives extra information on the relative numbers of hydrogen atoms in each group.

The n.m.r. spectrum of 1,4-dimethylbenzene (*p*-xylene) in Fig. 5.8, provides a good example of the way in which a structural analysis can be carried out. Thus, in addition to the peak at 0 δ from the added standard (TMS), two

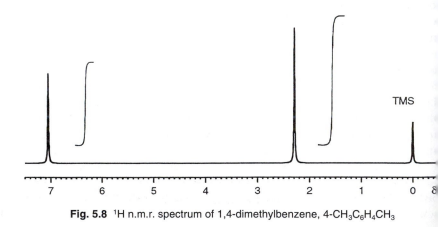

Fig. 5.8 ^1H n.m.r. spectrum of 1,4-dimethylbenzene, $4\text{-}CH_3C_6H_4CH_3$

bsorptions are detected: the ratio of the areas under the two absorptions is :3, and the peak at low-field (δ 7.05) is good evidence for the presence of romatic-ring hydrogen atoms. The other peak, at higher field, has a δ-value •f 2.3, characteristic of methyl groups attached to an arene. In this way the tructure is confirmed. Note that there are only *two* different types of hydro- en atom in the molecule and therefore two absorptions.

Although the information about the number and environment of a given ype of hydrogen atom in a molecule is undoubtedly of considerable assist- nce in structure determination, there is further helpful information which is •btained from n.m.r. spectra when they are recorded under conditions of high esolution.

•pin–spin couplings

Vhen care is taken to ensure that the magnetic field across the sample is homo- eneous, high-resolution conditions are achieved and some of the characteris- ic absorptions discussed previously are split into several components. We lready have an explanation for this because we started out by saying that vhen a simple bar magnet is placed in the magnetic field of a second magnet applied field), attractive and repulsive forces result. This situation also exists vhen two magnetically active nuclei are adjacent to one another. Clearly this ituation is common in many molecules. For example, Fig. 5.9 is a high reso- ution spectrum of ethanal which shows that the peak of area 3 (from CH_3, 2.21) is split into two (compare this with the low-resolution spectrum, 'ig. 5.4); the absorption is said to be a **doublet** and the two component lines ave the same intensity (1:1). On the other hand, the absorption from the HO hydrogen (δ 9.795) is split into a **quartet**, with relative intensities for he four lines of 1:3:3:1. The areas under the two *groups* of lines (i.e. the roups with different chemical shifts from the two types of hydrogen atom) re still in the ratio 3:1, as indicated by the integration trace.

The splittings can be explained by considering first the methyl-group ydrogen atoms. These experience an extra 'local' magnetic field due to the nagnetic moment of the CHO hydrogen atom. This proton's magnetic noment must be aligned either *with* or *against* the applied field, so that in the H$_3$–CH fragment, the CH_3 hydrogen atoms experience an extra field which an either *augment* or *oppose* the main field. Thus the resonance condition or the CH_3 hydrogen atoms can be achieved at two different external fields, vhich correspond to situations where the externally applied field is aug- nented by, or opposed by, the local field from the aldehydic hydrogen atom s shown. For a collection of CH_3CHO molecules some of the CHO-group ydrogen atoms will augment the main field, whereas in other ethanal mole- ules the magnetic moment of the CHO hydrogen atom will be opposed to he main field: absorptions will be seen for the CH_3 hydrogen atoms in both •f these possible environments and therefore there are two peaks for the reso- ance from the CH_3 hydrogen atoms. The CH_3 resonance is said to be 'split' •y the single hydrogen atom, and the distance between the two peaks of the loublet is referred to as the coupling (or spin–spin coupling, since the mag- etic moments are effectively produced by the spinning motion of the nuclei). Because both these states are almost equally populated (Section 5.1) the two •eaks have the same area.

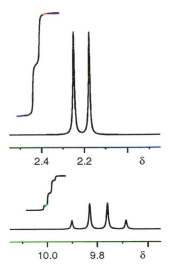

Fig. 5.9 High resolution spectrum of ethanal, CH_3CHO

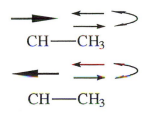

The CH_3 protons are in resonance; there are two possible magnetic fields generated by the CH proton which can reinforce or reduce the main field. These two situations result in the resonance of the CH_3 protons being split into a doublet.

A splitting into a doublet is characteristic of a group of equivalent hydrogen atoms (CH_3 in this example) split by *one* neighbouring hydrogen atom We can now derive the splitting patterns for larger numbers of neighbouring hydrogen atoms, as, for example, when the resonance from a hydrogen atom is split by an adjacent CH_2 or CH_3 group.

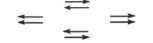

There are four ways of arranging the two nuclear spins of a CH_2 group. These produce three local magnetic fields which split the resonance from the adjacent hydrogen atoms into a **triplet**.

Splitting from two hydrogen atoms. Consider the spectrum from diethyl ether $CH_3CH_2OCH_2CH_3$ (Fig. 5.10), which has two main absorptions, from the hydrogen atoms of the CH_2 and CH_3 groups. The two sets of CH_3 hydrogen atoms (δ 1.15) will each experience local fields from adjacent $-CH_2-$ hydrogens, and we must therefore work out the possible 'arrangements' of the latter. Two CH_2 hydrogen nuclei may be both aligned with their magnetic moments in one direction (which we can represent ⇒), both in the other direction (⇐), or in opposite directions (⇆ or ⇄). This gives three possible local magnetic fields, of which, for a collection of molecules the arrangement with the magnets opposed (⇆ and ⇄) can be achieved in two possible ways and will be twice as common as the others Thus the splitting of the CH_3 resonance produces a 1:2:1 pattern, a triplet.

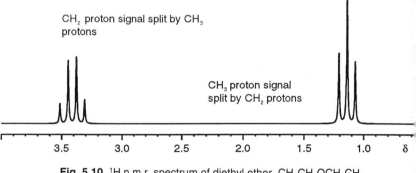

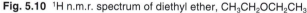

CH₂ proton signal split by CH₃ protons

CH₃ proton signal split by CH₂ protons

| 3.5 | 3.0 | 2.5 | 2.0 | 1.5 | 1.0 | δ |

Fig. 5.10 ^1H n.m.r. spectrum of diethyl ether, $CH_3CH_2OCH_2CH_3$

Splitting from three hydrogen atoms. In the two previous examples (CH_3CHO and $CH_3CH_2OCH_2CH_3$) we have to complete our understanding of the spectrum by realizing that the CHO hydrogen atom in CH_3CHO must be split by the CH_3 group and that the two sets of CH_2 hydrogen atoms in $CH_3CH_2OCH_2CH_3$ (at δ 3.4) must be split by the adjacent CH_3 hydrogen atoms. We therefore have to work out the possible local magnetic fields provided by the three magnetic moments of the protons of a CH_3 group. These are shown to the left.

There are four possible resultant magnetic fields, two of which can be achieved in three possible ways: thus a CH_3 group splits the resonance of neighbouring nucleus into a four-line pattern (**quartet**) with relative intensities 1:3:3:1, as confirmed by the splitting of the CH_2 absorption in Fig. 5.10 and of the CHO absorption in Fig. 5.9.

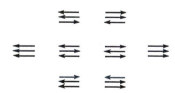

There are eight ways of arranging the three nuclear spins of a CH_3 group. This leads to four possible orientations of the local magnetic fields, and splits any coupled protons into a **quartet**.

More than three hydrogen atoms. In a fragment $-CH_2-CH-CH_2-$, in which the two CH_2 groups are equivalent (and have the same chemical shift) the single hydrogen atom will interact with four equivalent hydrogen atoms he resulting pattern is a 1:4:6:4:1 *quintet*. Five equivalent hydrogen atoms produce a splitting pattern of 1:5:10:10:5:1 (**sextet**), and six equivalent hydrogen atoms [e.g. in a $(CH_3)_2CH-$ group] split the single hydrogen atom's resonance into a 1:6:15:20:15:6:1 pattern (**septet**). The splitting patterns follow the coefficients of the terms in the binomial expansion and are conveniently expressed in the form of 'Pascal's Triangle':

Number of hydrogen atoms causing splitting	Splitting pattern produced (relative intensities of lines)
1	1 1
2	1 2 1
3	1 3 3 1
4	1 4 6 4 1
5	1 5 10 10 5 1
6	1 6 15 20 15 6 1

Summary

Before proceeding further, it is perhaps worthwhile being reminded of the various stages in the analysis of a complex n.m.r. spectrum by reference to that of CH_3CHO.

First, under low resolution conditions there are two absorptions, of relative area 3:1, from the CH_3 and the CHO hydrogens, respectively. These different groups have different chemical shifts (δ-values) which are typical of the chemical environment of the two types of hydrogen atom (Table 5.2).

Second, under high-resolution conditions, splitting can be seen. Interaction of the CH_3 hydrogen atoms with the CHO hydrogen atom means that the former resonance is a 1:1 doublet; interaction of the CHO hydrogen atom with the CH_3 hydrogen atoms means that the aldehydic hydrogen's resonance is a 1:3:3:1 quartet.

The size of the coupling, i.e. the separation between the peaks of each multiplet, is a measure of the energy of the interaction, and is the same for both resonances (see Fig. 5.9). The coupling is customarily quoted in frequency units ($\Delta\nu$ / Hz); it is, for a given compound, independent of the magnitudes of the characteristic radiofrequency and applied magnetic field of the n.m.r. spectrometer employed.

A word of explanation is necessary here since the measurement involves (usually) conversion from the δ-scale: the problem is to express a separation measured as $\Delta\delta$) as $\Delta\nu$ (Hz). From the definition of δ (page 59) we have:

$$\Delta\delta = \frac{\Delta B}{B_0} \times 10^6$$

and, from eqn 5.1,

$$\frac{\Delta B}{B_0} = \frac{\Delta\nu}{\nu}$$

so that

$$\Delta\nu = \frac{\nu.\Delta\delta}{10^6}$$

With a 400 MHz spectrometer (i.e. $v = 400 \times 10^6$ Hz), such as that used for many of the ^{1}H n.m.r. spectra shown in this book, 1δ unit is equivalent to 400 Hz. (For a 100 MHz spectrometer, 1δ unit equals 100 Hz.)

In the example CH_3CHO, the separation between the lines (for both the and the 1:3:3:1 patterns) is $0.0075\ \delta$ on a 400 MHz spectrometer. This corre ponds to 3 Hz (it is referred to as J_{HH}) and is always 3 Hz, no matter whi spectrometer is employed. The separation *measured as* $\Delta\delta$ is 0.03 on 100 MHz spectrometer.

Analysis of the splittings in the n.m.r. spectra of organic compounds usually fairly straightforward because appreciable coupling normally occu only between hydrogen atoms on *neighbouring* atoms (information about t alignment of one magnetic moment is transmitted to other nuclei through t bonds and the effect dies off rapidly with the number of intervening bonds This makes the technique extremely effective for distinguishing isomer alkyl structures (straight-chain and branched): see Section 5.3.

It must also be remembered that hydrogen atoms in exactly equivale environments in a molecule do not split each other. Thus, for CH_3CHO, f example, the methyl hydrogen atoms, while split by the CHO hydrogen, not split each other. This is because they are all identical and in resonan together (at the same chemical shift).

5.3   Examples of spectra showing spin-spin splittings

(a) Butanone (methyl ethyl ketone) $CH_3COCH_2CH_3$

The n.m.r. spectrum (Fig. 5.11) contains a single peak from the methyl grou adjacent to the carbonyl group, with the expected chemical shift ($\delta\ 2.15$; s Table 5.2). This peak is not split because there are no hydrogen atoms on t adjacent carbon atom.

In this example, as in many others, the splitting pattern matches the expected 1:2:1 and 1:3:3:1 relative intensities. Sometimes the peaks within such groups are slightly larger in the direction of the resonance of the group responsible for the splitting. The distortion (which can be ignored here) becomes more pronounced when chemical-shift differences become small. Examples are shown in Figures 5.14 and 5.15.

The other peaks are from the hydrogen atoms in the CH_2 group ($\delta\ 2.55$) ar the other CH_3 group ($\delta\ 1.05$); the CH_2 peak has the higher δ-value because the effect of the adjacent carbonyl group and the resonance appears as a 1:3:3 quartet because of the methene proton's interaction with the CH_3 group. T CH_3 resonance is split into a 1:2:1 triplet by the CH_2 hydrogen atoms (these tw multiplets form the characteristic pattern seen for an ethyl group).

As can be seen from Fig. 5.11, the integration trace still indicates the rel tive numbers of hydrogen atoms in the groups with different chemical shift

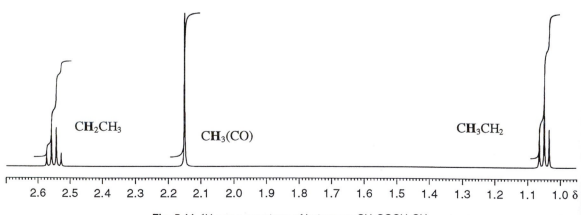

CH_2CH_3 $CH_3(CO)$ CH_3CH_2

2.6	2.5	2.4	2.3	2.2	2.1	2.0	1.9	1.8	1.7	1.6	1.5	1.4	1.3	1.2	1.1	1.0 δ

Fig. 5.11 ^{1}H n.m.r. spectrum of butanone, $CH_3COCH_2CH_3$

) 1,3-Dibromopropane, $BrCH_2CH_2CH_2Br$

e n.m.r. spectrum shown in Fig. 5.12 consists of two groups of resonances, e numbers of hydrogen atoms concerned being 2:1 (i.e. the relative total eas, as indicated by the height of superimposed steps in the integrated curve r each resonance). These are evidently the four outside and the two central ethylene hydrogen atoms, respectively, with the expected δ-values; the tside methylene groups absorb at higher δ value because of the inductive fect of the bromine atoms.

The central CH_2 resonance is split into a 1:4:6:4:1 quintet by the *four* ighbouring hydrogen atoms; the outside CH_2 groups each have one neigh-uring CH_2 group and hence appear as 1:2:1 triplets.

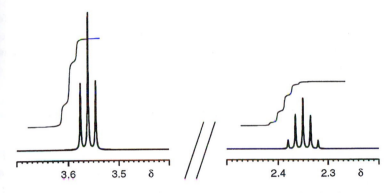

Fig. 5.12 ^{1}H n.m.r. spectrum of 1,3-dibromopropane, $BrCH_2CH_2CH_2Br$

) Bis-(1-methylethyl) ether (di-isopropyl ether), $(CH_3)_2CHOCH(CH_3)_2$

e spectrum (Fig. 5.13) indicates the presence of two different types of drogen atom, the numbers of each type being in the ratio 1:6. The methyl oup resonance (at δ 1.1) is split into a doublet, since each CH_3 group drogen atom experiences an interaction with the neighbouring single drogen atom. The CH resonance (δ 3.6) is split into a septet, characteristic interaction with six equivalent hydrogen atoms (see page 63).

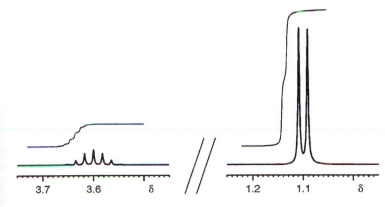

Fig. 5.13 ^{1}H n.m.r. spectrum of bis-(l-methylethyl) ether, $(CH_3)_2CHOCH(CH_3)_2$

For *p*-bromotoluene there are two pairs of chemically equivalent protons that experience slightly different magnetic fields due to the difference in electronic effects of Me and Br. The two pairs of protons therefore resonate at different chemical shifts.

(d) 1-Methyl-4-bromobenzene (p-bromotoluene), 4-CH₃C₆H₄Br

The spectrum (Fig. 5.14), recorded for a solution in CCl_4, shows clearly th resonances from aromatic (δ 7.4–7.0) and aliphatic (δ 2.3) hydrogen atom in the ratio 4:3. The methyl group resonance is not split and the aromat hydrogens now appear as two non-equivalent pairs: H_A, δ 7.35 and H_B, δ 7.(Here the signals for H_A and H_B are essentially both doublets because of th splitting for each hydrogen atom by the adjacent non-equivalent hydroge atom. This four-line pattern is typical of a 1,4-disubstituted benzene ring wi two different substituents.

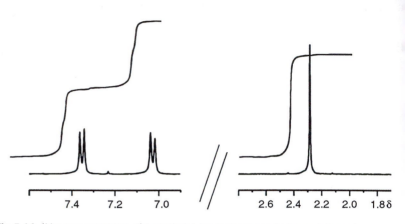

Fig.5.14 ¹H n.m.r. spectrum of 1-methyl-4-bromobenzene (*p*-bromotoluene), 4-CH₃C₆H₄Br

Not only does the observation of the splitting patterns in n.m.r. spectr indicate which groups are attached to others in an organic molecule, but also in certain cases, it is found that the magnitude of the splitting is informativ For example, in alkenes in which the hydrogen atoms are not all equivale (if they were, they would all have the same δ-value and hence have n observable splitting) then the following splittings are typical. The recognitio of these differences sometimes enables a choice to be made between variou possible isomeric structures. For example, in 3-phenylpropenoic acid (cir namic acid) the two alkenic protons resonate at δ 7.83 and 6.46, with coupling (J_{HH}) of 17 Hz: this must therefore be the *trans* (E) isomer.

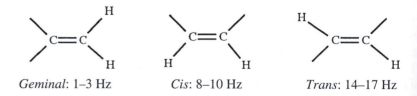

Geminal: 1–3 Hz   *Cis*: 8–10 Hz   *Trans*: 14–17 Hz

In some spectra, spin–spin couplings and even resonances which yo might expect to see are absent. This often happens when protons such as OI are exchangeable with those in the solvent. The observation of such behav iour is indicative of chemical exchange and represents a more advanced are of NMR study.

.4 Worked examples

t this stage you are encouraged to attempt to assign a structure that fits each pectrum shown in Figs 5.15–5.17, for each of which the molecular formula given.

These are followed by a discussion which gives the answers and some rief explanatory notes. Remember that the spectra offer three pieces of vital formation: the chemical shifts (related to the types of hydrogen atom resent), the integrated trace (the steps are proportional to the relative umbers of hydrogen atoms in each group), and the splittings (which depend n the number of hydrogen atoms in neighbouring groups).

iscussion

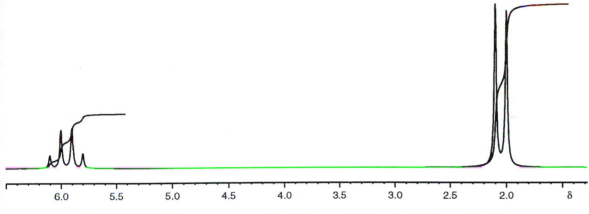

Fig. 5.15 60 MHz ^{1}H n.m.r. spectrum of worked example 5.1; molecular formula $C_2H_4Cl_2$

) Figure 5.15 is the spectrum of 1,1-dichloroethane, CH_3CHCl_2. There are early two types of hydrogen atom in the molecule, the number(s) in each roup being in the ratio of the integrated intensities (3:1). This strongly sugests the presence of CH_3 and CH groups, which is confirmed by the splitngs of 1:3:3:1 (of the CH absorption due to interaction with three equivalent ydrogen atoms) and 1:1 (the doublet from splitting of the methyl group bsorption by the single hydrogen atom). The high δ-value (5.95) for the ngle hydrogen atom reflects the negative inductive effect of the chlorine :oms.

i) Figure 5.16 is the spectrum of an aromatic compound, as judged from the bsorption at δ 7.2. The aliphatic part of the molecule has two types of hydroen atom, apparently in the ratio 1:6 (as deduced from the integrated trace). he high-field peak is typical of a C–CH_3 group, so that the part-structure H(CH_3)$_2$ may be suggested. This is confirmed by the splittings; the methyl ydrogen absorption is split into a doublet by interaction with one hydrogen :H) and the absorption of the single hydrogen is split by interaction with six quivalent hydrogen atoms to give a septet (1:6:15:20:15:6:1). Since the ratio f the number of aromatic hydrogen atoms to aliphatic hydrogen atoms is pproximately 5:7, the structure must be (1-methylethyl) benzene (isopropyl-enzene, cumene).

An expansion trace showing the aromatic proton region for (1-methylethyl) benzene is included in Fig. 5.16. The five aromatic protons of this molecule correspond to two chemically distinct pairs and a single proton. A highly complex pattern results because the chemical shift difference between the associated absorptions is small.

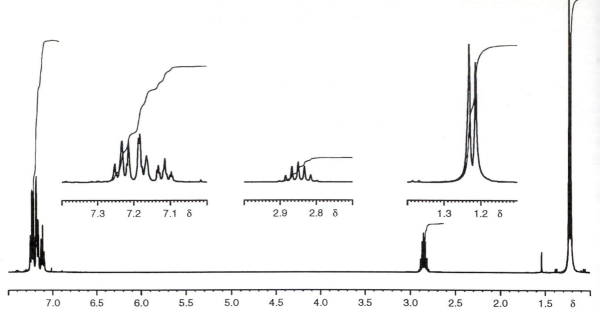

Fig. 5.16 400 MHz ^{1}H n.m.r. spectrum of worked example 5.2; molecular formula C_9H_{12}. Inset, expansions of peaks

(iii) From Fig. 5.17 we can conclude that there is a propyl group ($CH_3CH_2CH_2-$) present. Thus the high-field resonance at δ 0.95 (of relative intensity probably 3) is typical of a methyl group in an alkane, and there are two other absorptions with relative intensity 2: of these, the low-field (CH_2) peak is evidently split by a CH_2 group (to give a 1:2:1 triplet), and the splitting of the middle multiplet (evidently another CH_2) by CH_3 *and* CH_2 gives a 1:5:10:10:5:1 pattern. This is the spectrum of $CH_3CH_2CH_2NO_2$, although from the evidence so far presented the alternative structure $CH_3CH_2CH_2ONO$ cannot be ruled out. Distinction between these two would be made on the basis of the expected δ-values (e.g. from tables of data) for hydrogen atoms in alkyl nitrites ($-CH_2ONO$) and nitroalkanes ($-CH_2NO_2$), from other spectroscopic data (e.g. u.v., i.r.) or from chemical evidence.

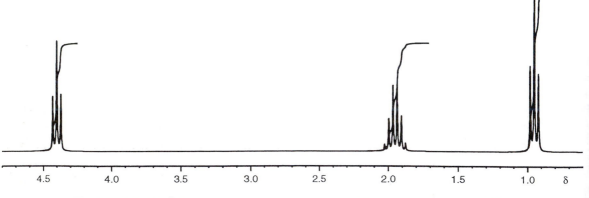

Fig. 5.17 200 MHz ^{1}H n.m.r. spectrum of worked example 5.3; molecular formula $C_3H_7NO_2$

Figures 5.15–5.17 clearly demonstrate the effectiveness of n.m.r. spectro-copy in distinguishing possible alkyl groups, e.g. $CH_2CH_2CH_3$ and $CH(CH_3)_2$.

5.5 Problems

.1 Identify the compound, of molecular formula C_3H_8O, whose n.m.r. spectrum is shown in Fig. 5.18.

.2 Identify the compound, of molecular formula $C_4H_8O_2$, whose n.m.r. spectrum is shown in Fig. 5.19.

.3 In the n.m.r. spectrum of 1,3-dibromopropane (Fig. 5.12) recorded on a 400 MHz spectrometer, the separation between the 1:2:1 peaks of the CH_3 resonance is 0.012 δ. Calculate the coupling $\Delta\nu$, (in Hz) between the methylene and methene hydrogen atoms in the molecule.

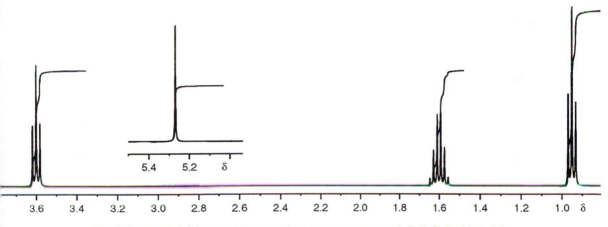

Fig. 5.18 400 MHz ^{1}H n.m.r. spectrum of an unknown compound, C_3H_8O: Problem 5.1

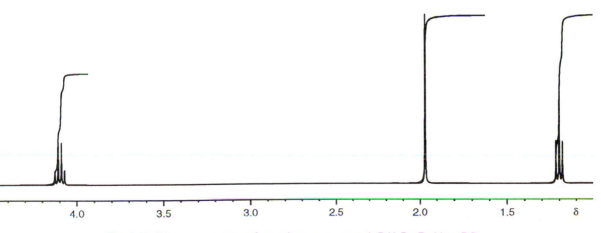

Fig. 5.19 ^{1}H n.m.r. spectrum of an unknown compound, $C_4H_8O_2$: Problem 5.2

5.4   The ^{1}H NMR spectrum of the compound whose mass spectrum was given in Problem 1.3 (page 16) and i.r. spectrum in Problem 3.4 (page 40) consists of a multiplet at δ 7.2 (area 5), a singlet at δ 5.1 (area 2), and a singlet at δ 1.95 (area 3). Does this confirm your assignment?

5.5   Predict the ^{1}H NMR spectra of the keto and enol forms of pentane-2,4-dione; the structures are given on page 51.

5.6   Pulsed n.m.r. spectrometers

We have already seen that n.m.r. spectroscopy can suffer from the limitation that detected signal strengths are very weak because the corresponding population differences are very small. While proton nuclei are the most readily detected, ^{13}C signals are 6400 times smaller due to their lower relative abundance and smaller magnetic moment. Consequently ^{13}C signals are much harder to detect.

Fortunately, these problems have been very effectively solved by the new generation of pulsed spectrometers now available for the detection of all n.m.r. active nuclei. These machines have revolutionized the scope of n.m.r., especially by allowing ready access to nuclei such as ^{13}C.

These instruments work by supplying a range of different **frequencies** instantaneously to the sample which is located in a magnetic field of constant value. This is achieved by irradiating the sample with a short **pulse** of electromagnetic radiation such that the applied frequencies are in the region expected to contain the resonances in the selected sample. For example, if the spectrometer employed to record the ^{13}C spectra has a magnetic field of 1.41 Tesla, the frequency of the pulse is 15 MHz (i.e. the appropriate resonance frequency for ^{13}C in this field) and the spread of frequencies about this value ($\Delta\nu$) is typically 4 kHz.

The effect of the radiation is to excite the ^{13}C nuclei whose resonance frequency lies within the range $\Delta\nu$. After the pulse, these ^{13}C nuclei return to their equilibrium state within a few seconds but during that time a signal can be detected which contains information about all the frequencies that have been excited (i.e. the resonance frequencies). The measured analog signal is converted into a digital value and stored in computer memory. Repeating the pulse-additive-data-storage process rapidly allows the detected signals to be added together to increase the signal strength relative to the background noise level before conversion into the n.m.r. spectrum. Because this process enhances the signal-to-noise ratio it improves the quality of the final spectrum and thereby allows the measurement of very weak signals.

A technique referred to as the Fourier Transform method is employed to convert the signal stored on the computer into the final spectrum. Put simply, this process separates the complex mixture of frequencies detected after the pulse into its component frequencies. The resulting spectrum of different frequencies at a fixed field is plotted in the same way as described earlier, i.e. chemical shifts on the δ scale, relative to TMS.

N.m.r. from other nuclei

We have already seen that besides ^{1}H, other nuclei have nuclear magnetic moments: these include ^{2}H (deuterium), ^{11}B, ^{13}C, ^{14}N, ^{19}F, ^{31}P, ^{35}Cl, ^{37}Cl, ^{79}Br, and ^{81}Br. These nuclei also interact with a magnetic field and therefore exhibit their own resonance condition. Furthermore, these nuclei will interact with other magnetically active nuclei and may give rise to splittings in a similar way to the ^{1}H–^{1}H couplings described earlier. However, the ease of measuring these splittings depends on the nucleus, and it is rarely possible to observe splittings from ^{14}N or the chlorine and bromine isotopes.

Importantly, the value of the magnetic moment μ is a characteristic of the nucleus that is being examined. Because the values of μ differ substantially from one nucleus to the next, the resonance frequencies for each nucleus are normally well separated. This is illustrated in Fig. 5.20 and confirms that, for example, the resonance frequency of ^{19}F atoms in fluorine-containing molecules can be detected at a frequency which is 94% of the ^{1}H frequency. By tuning the circuit of an n.m.r. probe for resonance it is possible to affect, and detect, only selected values of ν, in other words, probe the magnetically active nuclei in the sample selectively. This process is similar to tuning into a radio station on your stereo and means that several n.m.r. experiments must be completed before all the magnetically active nuclei in a sample have been monitored.

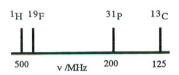

Fig. 5.20 Resonance frequencies for different nuclei in an external field of 11.74T.

5.7   Carbon-13 n.m.r. spectroscopy

It is particularly desirable to be able to record n.m.r. signals from the very small proportion of carbon nuclei (^{13}C) which possess magnetic moments (you will recall that ^{12}C does not have a magnetic moment). This enables information to be gained directly about a molecule's carbon skeleton, with the presence of functional groups such as CO or CN being easy to recognize.

Carbon-13 spectra do not normally exhibit ^{13}C–^{13}C splittings because there is a very low natural abundance of ^{13}C, and the percentage of molecules which possess two adjacent ^{13}C atoms is very small. However, there will be interactions between the ^{13}C nuclei and the magnetic moments of nearby protons in the molecule, which will cause splittings (exactly as for ^{1}H–^{1}H splittings). The resulting patterns are often quite complicated, however, because a particular ^{13}C nucleus may show observable splittings not only from protons attached directly to it but also from protons further away.

Fortunately, it is possible to simplify the spectra and remove *all* these splittings by simultaneous *decoupling* of all the protons in the sample. To achieve this the sample is irradiated with an extra range of frequencies such that all the proton's resonances are excited while the ^{13}C signals are recorded (see Section 5.6). The absorption of this extra radiation means that each proton no longer provides a static local magnetic field for neighbouring carbon nuclei but rapidly changes from one spin state to the other with the result that the effect which normally causes the splittings is averaged and lost. Fig. 5.21 shows a typical fully decoupled ^{13}C n.m.r. spectrum which has been recorded in this way: this is for butanone, $CH_3COCH_2CH_3$, whose ^{1}H n.m.r. spectrum is shown on page 64. The spectrum was obtained by adding signals from 32 pulse-collect measurements in two minutes.

Each of the carbon atoms in butanone gives an absorption which is a single sharp peak, with a different chemical shift; it is the range of shifts obtained, plus the simplification gained by decoupling, that is particularly

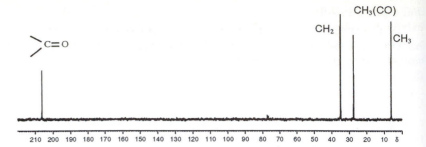

Fig. 5.21 ^{13}C n.m.r. spectrum (fully decoupled) of butanone, $CH_3COCH_2CH_3$

useful in the study of large molecules such as proteins, alkaloids, and steroids. For example, the steroid cholesterol-acetate, whose structure is shown in Fig. 5.22, has a broad, complex, and relatively uninformative proton n.m.r. spectrum (the different alkyl fragments have very similar chemical shifts and splittings from adjacent protons, causing considerable overlap). In contrast, the fully decoupled ^{13}C spectrum shown in Fig. 5.23 shows significant differences in the carbon chemical shifts, and resonances from most of the 29 carbon atoms can be recognized.

In the case of ^{13}C spectra, unless special precautions are taken the peaks will have heights which are not strictly proportional to the relative numbers of different carbons in the molecule and integration is not normal. However, the peak heights often serve as a useful guide to abundance.

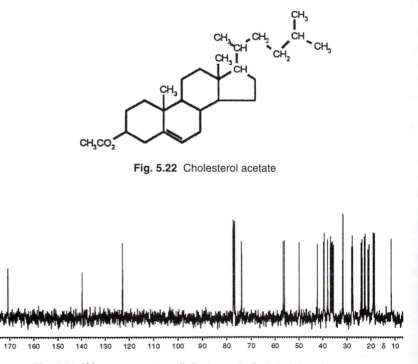

Fig. 5.22 Cholesterol acetate

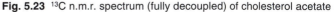

Fig. 5.23 ^{13}C n.m.r. spectrum (fully decoupled) of cholesterol acetate

^{13}C chemical shifts

^{13}C chemical shifts, like ^{1}H n.m.r. shifts, are also found to be diagnostic of structure, and typical ranges of ^{13}C shifts for different types of compound and functional groups are given in Fig. 5.24.

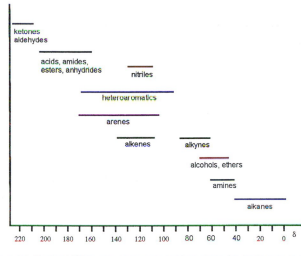

Fig. 5.24 Typical ^{13}C n.m.r. chemical shift regions for indicated groups

Carbonyl groups, and carbons in aromatic rings, alkene, and alkyl fragments can usually be readily distinguished. For example, in the spectrum shown in Fig. 5.21, the absorption at δ 207 is typical of the carbonyl group in the ketone. Likewise, the C=C carbon atoms in cholesterol acetate are characterized by the peaks at δ 140 and δ 123 in Fig. 5.23.

At this stage you are encouraged to attempt to identify the compounds whose fully decoupled ^{13}C n.m.r. spectra are shown in Figs 5.25 and 5.26 and for which the appropriate formulae are provided.

In the first of these there are only two separate types of ^{13}C atom (with one peak approximately twice as high as the other). This indicates that of the three carbon atoms, two are in identical environments: the structure $ICH_2CH_2CH_2I$ would clearly fit, whereas the alternatives $CH_3CH_2CHI_2$ or CH_3CHICH_2I would each be expected to have three separate resonances.

In the second molecule, which clearly contains three carbon atoms in different environments, the peak at δ 172 is typical of a carbonyl group in, for example, an ester or acid: on the basis of this spectrum and the formula, the structure $CH_3C(O)OCH_3$ is suggested.

Carbon–proton multiplicity determination

The main disadvantage of the otherwise helpful *full* decoupling procedure is that vital structural information which would result from the observation of C−H splittings is lost. Fortunately, it is also possible to record a ^{13}C n.m.r. spectrum employing a partial decoupling procedure such that the splitting pattern between each carbon and the attached protons is effectively retained but further splittings are removed. This is called 'off-resonance' decoupling.

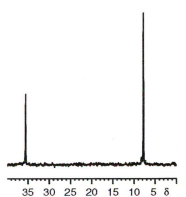

Fig. 5.25 ^{13}C n.m.r. spectrum (fully decoupled) of an unknown compound $C_3H_6I_2$

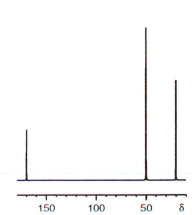

Fig. 5.26 ^{13}C n.m.r. spectrum (fully decoupled) of an unknown compound $C_3H_6O_2$

The number of lines in the patterns and their relative heights are governed by the same simple rules formulated for ^1H–^1H splittings, and Pascal's triangle (see page 63) again provides the key guidelines. For example, the spectrum of butanone recorded in this way is shown in Fig. 5.27, which should be compared with the fully decoupled spectrum in Fig. 5.21. Note that the methyl-group carbon absorptions at δ 28 and δ 7 are now split into quartets which are approximately 1:3:3:1 patterns, indicative of three attached protons in each group; the methylene carbon absorption at δ 36 becomes essentially a triplet (1:2:1) pattern, whereas the absorption at δ 207, from C=O, has no splitting since this carbon has no attached protons. The off-resonance decoupled spectrum of the example in Fig. 5.26 also has quartets at δ 50 and δ 20, which should help confirm your structure.

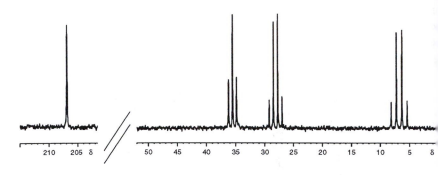

Fig. 5.27 ^{13}C n.m.r. spectrum (off-resonance decoupled) of butanone

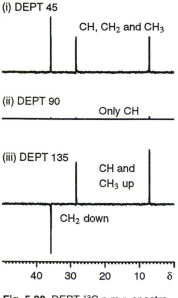

(i) DEPT 45

CH, CH$_2$ and CH$_3$

(ii) DEPT 90

Only CH

(iii) DEPT 135

CH and CH$_3$ up

CH$_2$ down

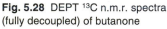

Fig. 5.28 DEPT ^{13}C n.m.r. spectra (fully decoupled) of butanone

This approach has now been replaced by a series of high-sensitivity experiments which enable carbon–proton multiplicity to be assigned by inspection, via the so called 'spectral editing' method. Here, after a complex series of radiofrequency pulses (the DEPT sequence) is applied to the sample, the detected signals are controlled by the carbon–proton multiplicity; some peaks appear as a positive-going signals, above the baseline, and some as negative-going signals, below the baseline. Fig. 5.28 illustrates the DEPT spectra of butanone. A proton decoupled ^{13}C spectrum which shows signals from all carbons is normally recorded first (Fig. 5.21). Then ^1H decoupled DEPT-45 (i), DEPT-90 (ii), and DEPT-135 (iii) spectra are collected. These DEPT spectra only contain signals from carbons with protons directly attached and are easier to measure than 'normal' spectra. In case (i) all the carbon signals arising from groups with protons attached appear as positive-going signals, while in case (ii) the only visible signals arise from methine carbons, and in case (iii) methyl and methine signals appear as positive peaks while the methylene signals are negative. When these four spectra are examined together the unambiguous assignment of carbon resonances to functional groups can normally be achieved.

5.8   Conclusion

The development of n.m.r. spectroscopy is continuing in many exciting ways. The vast amount of detailed information to be obtained from ^{13}C spectra (from shifts, coupled, and decoupled spectra) has led to the manufacture of pulsed spectrometers of enormous sophistication that can be used to examine the wide range of nuclei with magnetic moments (^2H, ^{18}O, ^{19}F, ^{31}P, ^{33}S, etc.). These changes have also facilitated the linking of resonances in two dimensions and now, for example, ^1H and ^{13}C spectra can be connected in a single plot. These new techniques lead to incredible versatility in the type of analytical and research problems tackled by n.m.r.

The development with perhaps the greatest implications of all is 'magnetic resonance imaging' or 'whole-body' n.m.r. In this approach, a much larger sample (e.g. a human body or limb) is placed in the magnetic field and a signal obtained from, for example, ^1H nuclei in water molecules in human cells. Signals from healthy and diseased tissues can often be distinguished and, as with X-rays, an image can be obtained. These studies, compared with related applications of X-rays, have the important advantage that the radiation employed is harmless. The scan illustrated in Fig. 5.29 is the MRI image (from ^1H nuclei) of the neck region of a human spinal column.

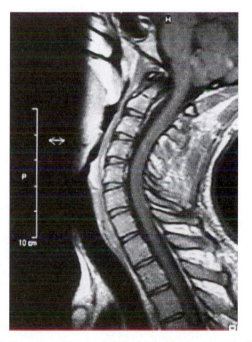

Fig. 5.29 Magnetic Resonance Image (MRI) of part of a human spinal column, one of a series of scans which make up a 3D-image.

Further Reading

1. L. M. Harwood and T. D. W. Claridge, *Introduction to Organic Spectroscopy*, Oxford Chemistry Primers, Oxford University Press, Oxford, 1997.
2. P. J. Hore, *Nuclear Magnetic Resonance*, Oxford Chemistry Primers, Oxford University Press, Oxford, 1995.

6 X-ray diffraction

The diffraction techniques are quite different from the spectroscopic methods so far discussed. Whereas the latter are based on the absorption of certain wavelengths (and energies) from radiation with a range of wavelengths, diffraction techniques employ radiation with a single wavelength. i.e. *monochromatic* radiation.

X-ray diffraction occurs when a monochromatic beam of X-radiation interacts with matter and is *scattered* in different directions, with no absorption of energy. Similarly, a beam of neutrons or electrons with well-defined wavelength can be scattered to give typical diffraction patterns.

The basis of the application of diffraction techniques in chemical problems is to use ions or molecules as diffraction gratings and then to determine, from the observed diffraction phenomena, the spacings between ions in a crystal or between the atoms which constitute molecules.

6.1   Introduction to the X-ray diffraction method

The first significant experiments were carried out at the beginning of the twentieth century when it was realized both that X-rays have wave properties and that crystals consist of regular arrays of atoms or ions. It was demonstrated by von Laue that a crystal lattice behaves as a grating so that it is possible to generate a diffraction pattern from a crystal; for his first experiments he used an ionic crystal, a beam of X-rays (with a range of wavelengths) and a photographic plate (to detect the scattered X-rays). A regular pattern of spots appeared on the plate, giving a clear indication of the success of the experiment. As we shall see, diffraction phenomena can be observed if the wavelength of the radiation is of the same order of magnitude as the 'repeat distance' of the atoms or ions in a crystal; X-rays (but not visible light) fulfill this condition.

The method has been developed to provide a means for determining the exact positions of ions in an ionic crystal lattice and of atoms within a molecule — that is, for determining accurate values for bond angles and bond lengths, even in extremely complicated molecules like proteins and enzymes.

The apparatus

X-rays are produced when a beam of accelerated electrons strikes a metal target. An inner electron from an atom in the metal is ejected, an outer electron drops down to fill the vacancy created, and the emitted radiation, of precise energy, frequency, and hence wavelength, is in the X-ray region. Since various electronic transitions are possible, the resultant beam at this stage contains X-rays of several different energies (wavelengths). Figure 6.1 for example, shows the intensity of X-radiation, as a function λ, emitted from a copper target. The K_α line, which corresponds to the energy emitted when

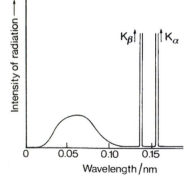

Fig. 6.1 Intensity of X-radiation at different wavelengths emitted from a copper target

an electron undergoes a transition from the *L* shell to the *K* shell ($2p \rightarrow 1s$ in terms of orbitals), has $\lambda = 0.154$ nm. A sheet of nickel proves to be a good filter for all the wavelengths except the K_α line (all the X-rays of wavelength less than the 'absorption edge' of 0.149 nm are absorbed: they are of high enough energy to remove completely a *K*-shell electron from a nickel atom) so the combination of the two metals used like this provides a *monochromatic beam* of radiation.

The monochromatic beam of X-rays is incident on a solid sample (single crystal or powder) of the material under investigation. A powder contains many very small crystals, in a variety of different orientations, whereas for a single crystal only one orientation of the solid can be considered at a time.

Detection of the resultant X-ray beam is usually achieved by surrounding the sample with a photographic film: where X-rays strike the film it becomes darkened, and the film is subsequently developed to yield the diffraction pattern. Figures 6.2a and b show the experimental arrangement for taking a powder 'photograph'.

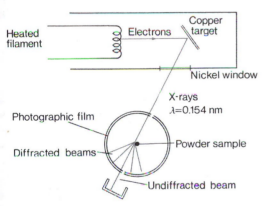

Fig. 6.2a Basic features of an X-ray camera for use with powdered samples

Fig. 6.2b Arrangement of the film for a powder photograph: the resulting 'lines' can be seen

Figures 6.3 and 6.4 are the thin strips of film (opened out) recording the diffraction patterns from two powdered metals (molybdenum and copper, respectively). Figure 6.5 is the diffraction pattern from powdered sodium chloride. The photographs indicate that the X-rays, on striking the powders, become diffracted into a series of well-defined cones. The origin of this phenomenon will be discussed in the next section.

If a *single crystal* is employed, it is usually mounted at the centre of a cylindrical film of somewhat greater depth than that used in the powder method. The crystal is arranged with one of its major axes vertical. The diffraction pattern is then recorded, often with simultaneous rotation of the crystal about the axis. A typical *single crystal photograph* shows several *layers* of spots; Fig. 6.6 is an X-ray single crystal rotation photograph of sodium chloride. The derivation of information from the photographs will follow the explanation of the origin of the patterns.

(1, 1, 0)   (2, 0, 0)   (2, 1, 1)   (2, 2, 0)

Fig. 6.3 X-ray diffraction photograph from a powder sample of molybdenum

(1, 1, 1) (2, 0, 0)   (2, 2, 0) (3, 1, 1) (2, 2, 2)

Fig. 6.4 X-ray diffraction photograph from a powder sample of copper

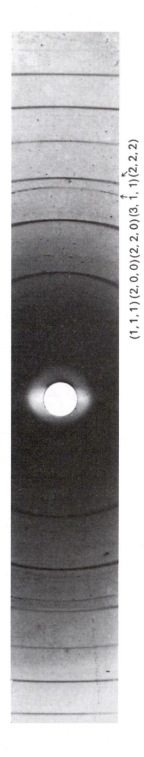

(1, 1, 1) (2, 0, 0)(2, 2, 0)(3, 1, 1)(2, 2, 2)

Fig. 6.5 X-ray photograph from a powder sample of sodium chloride

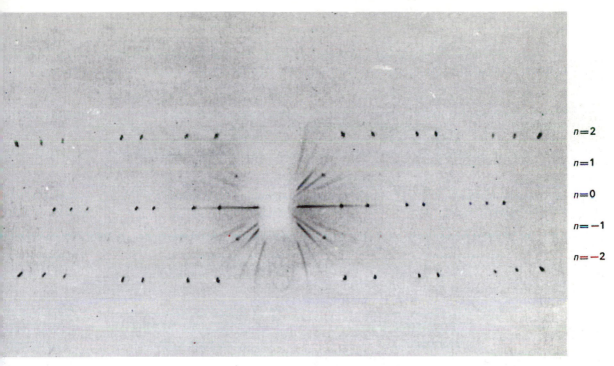

Fig. 6.6 Single-crystal rotation X-ray photograph

The Bragg equation

A more detailed understanding of X-ray diffraction and of the exact requirements for the appearance of intensity maxima were presented by W. L. Bragg in 1912. He realized that when X-rays impinge on a crystal some are reflected from the atoms in the top layer, whereas others penetrate this layer and are reflected off the next layer, and so on. Analysis shows that the resultant reflected rays are only in phase for certain angles of incidence of the X-rays upon the crystal. This is illustrated in Fig. 6.7, which shows the path difference between the two rays (one reflected off the top layer, the other reflected off the second layer) when they arrive at the at the detector. For these two rays to combine (i.e. to reinforce each other) they must be completely in phase: that is, their path difference ($2d\sin\theta$) must be a whole number of wavelengths ($n\lambda$) where d is the separation between the planes and λ is the wavelength of the X-rays. This is the **Bragg Equation** (6.1) and it correctly predicts that a reflected beam will be observed (i.e. that there is *constructive* interference) only for certain angles of incidence of the X-ray beam on the crystal; at other angles of incidence the rays from the different layers will be partly or completely out of phase (*destructive interference*).

$$n\lambda = 2d\sin\theta \qquad (6.1)$$

For example, if $\lambda = 0.154$ nm and $d = 0.2$ nm, then as θ (the angle of incidence) is steadily increased, reflection will first occur when $n = 1$ and $\sin\theta = 0.385$, i.e. when θ is approximately 23°. The condition may also be satisfied for $n = 2,3,4$ etc., and hence for higher values of θ, but we will normally be concerned with the *first order* ($n = 1$) reflections.

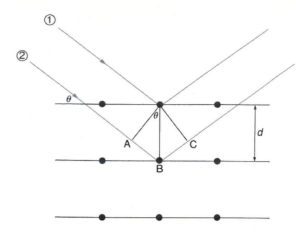

Fig. 6.7 Reflection of X-rays from the first and second rows of atoms (ions) in a solid lattice; the path difference between the reflected rays is (AB + BC) = $2d\sin\theta$, where θ is the angle of incidence and d is the distance between the planes

Figure 6.8 illustrates one particular small crystal (crystallite) in a *powder* orientated with its surface plane of atoms at an angle θ to the beam. If this value of θ fulfils the Bragg equation for the particular values of λ and θ in the experiment (since there are many crystallites in the powder, this will be true for some of them), then the beam is reflected to the film. There will be other crystallites each making the angle shown, which means that a cone of diffracted X-radiation will be produced, darkening the film at the points indicated (see also Figs 6.2–6.5). A *series* of cones is produced because there are various possible spacings in the crystal (with different values of d) which satisfy equation 6.1 for different values of θ.

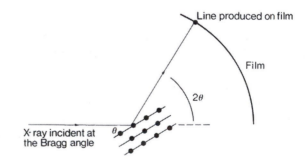

Fig. 6.8 Production of a 'line' on the photographic film from diffraction of X-rays incident at the Bragg angle

When a *single crystal* is used, with one axis vertical, then the regular inter atomic spacing along this axis behaves as a simple diffraction grating (Fig. 6.9). Thus, the layers of lines observed (see Fig. 6.6) are simply the reflections which satisfy the equation for λ, d (the vertical spacing, Fig. 6.9) and $n = 1, 2$, *etc.*

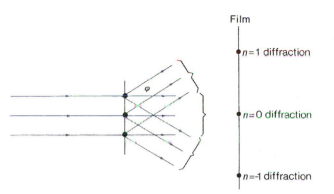

Film

● $n=1$ diffraction

● $n=0$ diffraction

● $n=-1$ diffraction

Fig. 6.9 Diagrammatic representation of diffraction by a single crystal mounted with one axis vertical; the diffractions with $n = 1$, 0, etc. correspond to the layers of lines clearly visible in Fig. 6.6

The layer lines themselves are clearly not continuous, because for some points on the layer lines there will be destructive interference between the reflections which satisfy eqn 6.2 for the vertical spacing and reflections from other planes in the crystal. To understand this and to appreciate the relationship between the possible spacings in a crystal (and hence all the possible values of d) and the structure of the lattice, some familiarity with crystal types is needed.

$$n\lambda = d\sin\varphi \qquad (6.2)$$

6.2 Crystallography

Unit cells and crystal systems

A crystal consists of a repeating **unit cell** of atoms or ions, in three dimensions. The unit cell is characterized by the length of each side (a, b, c) and the angles between the three sides (α, β, γ). Figure 6.10 shows two of the possible different types of unit cell or **crystal system.** The symmetry of each of these allows three-dimensional structures (crystals) to be built up from the many building blocks (unit cells). The planes referred to earlier are sheets in the crystal containing a high density of lattice points (atoms or ions), observed as the external faces of a crystal.

Within the unit cell there are various allowed arrangements of atoms or ions which still preserve the overall symmetry. Figure 6.11 shows the three possibilities for a **cubic** unit cell (i.e. with equal sides, all angles 90°); these are the **primitive** (or **simple**) cubic, **body-centered** cubic, and **face-centered** cubic systems.

You will probably find it helpful inspect 'ball-and-stick' models of unit cells and fragments of lattices in order to appreciate the various possible planes of atoms or ions in the different types of crystal. You may also find it helpful that a face-centred cubic lattice can equally well be described as a cubic-close-packed arrangement; this common structure arises if equivalent spheres are packed in one layer, the next layer is added, and then the third layer is added at positions not corresponding to those of the first layer, but in the alternative positions; the next layer corresponds to the first, and the pattern is repeated.

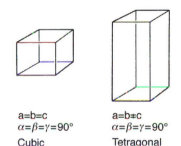

$a=b=c$
$\alpha=\beta=\gamma=90°$
Cubic

$a=b\neq c$
$\alpha=\beta=\gamma=90°$
Tetragonal

Fig. 6.10 Two crystal systems

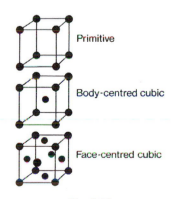

Primitive

Body-centred cubic

Face-centred cubic

Fig. 6.11

Fig. 6.12 Unit cell of sodium chloride, NaCl. One type of ion (Na$^+$ or Cl$^-$) can be seen at the corners of the cube and at the centre of each face: the other ion occupies vacancies created in this lattice and itself has a face-centred cubic structure. (Model by Crystal Structures Ltd.)

Lattices in solids

The lattices to be studied may be of several types.

(i)   *Atoms (metals or alloys).* For example, metallic caesium exists in a body-centered cubic lattice, as does α-iron, whereas copper has a face-centered cubic pattern. There is only one type of atom in the unit cell.

(ii) *Ionic crystals.* These contain two or more different types of ion (e.g. caesium chloride, sodium chloride), and they can be characterized in the same way as the simple lattices. For example, caesium chloride consists of a simple cubic array of Cs$^+$ ions with Cl$^-$ ions at the body centre and *vice versa.* The arrangement is referred to as two inter-penetrating primitive cubic lattices. For sodium chloride the crystal structure is of two inter-penetrating face-centred cubic lattices: the complete unit cell is shown in Fig. 6.12. Remember that the use of small balls and 'bonds' for models is essentially for convenience in visualizing the planes and unit cells. Close packing of larger space-filling spheres produces a more realistic model in which the electron clouds of neighbouring ions are seen to be in contact.

(iii) *Covalent molecules.* Covalent compounds also form crystals, with *molecules* at the lattice points in the unit cells, as for atoms and ions.

Planes in the crystal

A shorthand procedure is used for describing a particular plane in a crystal. (**Miller Indices**). Their definition can be illustrated for the three-dimensional lattice illustrated in cross-section in Fig. 6.13; the *x* and *y* axes are indicated, with the *z*-axis coming out of the paper. Each dot then represents a vertical column of atoms. The three lines indicated are three planes which we wish to describe.

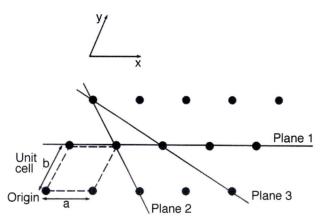

Fig. 6.13 Representation of different planes within a crystal

The procedure is as follows:
(i)   choose an origin;
(ii) read off the intercepts along the *x, y* and *z* axes in terms of the unit spacings in the crystal (*a* along the *x*-axis, *b* along the *y*-axis, *c* along the *z*-axis); the intercept is infinity if the plane does not cut the axis concerned;

iii) take reciprocals, dropping any reference to a, b, and c; and

iv) if the reciprocals for a given plane include a fraction or fractions, express the reciprocals as their simple integral ratio, leaving Miller Indices, which are then referred to as h,k,l values.

Step (i) Plane	(ii) Intercepts			(iii) Reciprocals			(iv) Miller Indices h,k,l
(1)	∞a	$1b$,	∞c	$1/\infty$	$1/1$	$1/\infty$	(0,1,0)
(2)	$2a$,	$2b$,	∞c	$1/2$	$1/2$	$1/\infty$	(1,1,0)
(3)	$4a$,	$2b$,	∞c	$1/4$	$1/2$	$1/\infty$	(1,2,0)

Thus plane (1) and *all those parallel to it* are referred to as (0,1,0) planes; similarly plane (2) and those parallel to it are called (1,1,0) planes. You should check that it does not matter where the origin is taken.

To recapitulate: a, b, and c are properties of the unit cell, as are the angles between axes; h, k and l are integers used to describe any particular plane in the crystal and, as will be seen, they are used to calculate the distance between the planes (i.e. d in the Bragg equation). Incidentally, a three-dimensional lattice can cleave (or grow) along any of these planes, which accounts for the external forms of crystalline compounds.

At this stage you are encouraged to classify some planes to become familiar with the nomenclature. Given the planes and axis system in Fig. 6.14, (i) what are the Miller Indices for each plane, and (ii) can you draw the (0,0,2) plane?

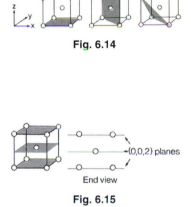

Fig. 6.14

To visualize a plane running through the crystal it is perhaps best to consider several unit cells together, or even better, to refer to a model. The answers for the planes shown are from left to right, (0,0,1), (1,1,0), and (1,1,1), respectively. Note that the (1,0,0) and (0,1,0) planes can simply relate to the (0,0,1) plane, i.e. they are all the appropriate 'ends' of the cube. Similarly, (1,0,1) and (0,1,1) will be diagonal planes (like 1,1,0).

The (0,0,2) planes are as shown in Fig. 6.15. To check this, note that the intercepts are ∞a, ∞b, $\frac{1}{2}c$, with *reciprocals* (0,0,2).

Fig. 6.15

It is often necessary to describe the perpendicular distance d_{hkl} between parallel planes (h,k,l) of a unit cell with lengths a, b, and c. This is the interplanar spacing referred to in the Bragg equation. With some relatively simple geometry it can be shown that for an **orthogonal** crystal system (a unit cell with all angles 90°):

$$\frac{1}{d^2_{hkl}} = \frac{h^2}{a^2} + \frac{k^2}{b^2} + \frac{l^2}{c^2} \tag{6.3}$$

and for a **cubic** system, $d_{hkl} = a/\sqrt{(h^2 + k^2 + l^2)}$ (eqn 6.4). For example, the distance between 002 planes is $a/2$, as expected.

6.3   Determination of structure

Unit cell type and dimensions

X-ray data from a given ionic crystal — in terms of the particular value of θ at which reflections are observed — can be used to determine the type of unit

$$n\lambda = 2d\sin\theta \qquad \text{[eqn (6.1)]}$$

$$\therefore \sin^2\theta = \frac{n^2\lambda^2}{4d^2}$$

$$\therefore \sin^2\theta = \frac{n^2\lambda^2}{4a^2}(h^2 + k^2 + l^2) \quad (6.5)$$

cell in the compound. The Bragg equation [for the allowed values of θ (angle of incidence) for given λ and d] is combined with the expression for the possible values of d in terms of the cell dimensions (a,b,c) of the particular structure considered. For a cubic lattice, for which the spacing d_{hkl} between any set of planes (h,k,l) is given by eqn 6.4, the resulting expression which gives the angles of reflection from different planes is derived as shown, where a is the length of the cell side (eqn. 6.5).

There are many lines in the diffraction pattern (i.e. at different values of θ) owing to the various possible values of d. Since for any set of planes h,k, and l must all be integers, then so must ($h^2 + k^2 + l^2$); the possible values of h, k, l and ($h^2 + k^2 + l^2$) are as follows:

plane

h,k,l	1,0,0	1,1,0	1,1,1	2,0,0	2,1,0	2,1,1	2,2,0	$\begin{cases} 2,2,1 \\ 3,0,0 \end{cases}$	3,1,0
($h^2 + k^2 + l^2$)	1	2	3	4	5	6	8	9	10

Thus for a cubic lattice, the increasing values of $\sin^2\theta$ should be related to each other as are the increasing simple integers [the smallest value of $\sin^2\theta$ being the reflection from the (1,0,0) planes, the next from the (1,1,0) planes, etc.] but with no line corresponding to the integer 7 because no combination of the squares of integers gives this number. Similarly it can be shown that there should be no lines for integers 15, 23, 28. The pattern obtained from a simple cubic lattice confirms this analysis, and any solid which gives such a pattern (i.e. with 7th, 15th etc., lines missing) is known to have a cubic lattice. Further, if λ is known, values of $\sin^2\theta$ for the $n = 1$ reflections lead to a measurement of a, the length of the side of the unit cell. The process just described is called **indexing** a powder photograph (that is, deriving the shape and size of the unit cell).

In the single crystal method, measurements with each axis vertical in turn lead to determination of the unit cell length along each of the axes (from the separation of the layer lines, see page 81). Again, comparison with patterns from compounds whose structures are known can be helpful.

The Bravais lattice

The type of Bravais lattice (e.g. body- or face-centred) can also be readily obtained from the diffraction pattern. First, consider a body-centred cubic structure and try to visualize what happens to the expected reflections from the (1,0,0) [and the (0,0,1) (0,1,0) planes]. For the (1,0,0) planes viewed end-on, it can be seen that the reflection which satisfies the Bragg equation (i.e. with distance d_{100}) now has an extra beam superimposed (Fig. 6.16). This is the beam reflected from the atoms at the centres of the unit cells; as shown this extra beam is completely *out of phase* with the beams reflected from the (1,0,0) planes (the path difference for the extra beam compared to the others is $\lambda/2$ and therefore there will be destructive interference; the reflection is now absent). This is referred to as a **systematic absence**.

The (2,0,0) (0,2,0), and (0,0,2) reflections (with separation d_{200}) will be present, the reflections being in phase (remember that d is now half that for the 1,0,0 planes); a diffracted beam at the corresponding θ is observed. As

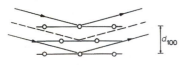

Fig. 6.16

Fig. 6.17 demonstrates, the (1,1,0) reflection should be *present* [there being no atoms in between the (1,1,0) planes] but the (1,1,1) plane reflections are absent (the planes have atoms in between).

For a face-centred cubic structure, the (1,0,0) reflections are absent, as too are the (1,1,0) reflections, but not, in this case, the (1,1,1) type (you are recommended to check that this is the case by reference to models). The following diagram summarizes the observed reflections:

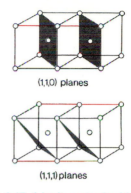

(1,1,0) planes

(1,1,1) planes

Fig. 6.17 A body-centred cubic structure showing planes with interstitial ions (1,1,1) and planes without interstitial ions (1,1,0)

Cell type	(h,k,l)	1,0,0	1,1,0	1,1,1	2,0,0	2,1,0	2,1,1	2,2,0
Primitive cubic		✓	✓	✓	✓	✓	✓	✓
Body-centred cubic			✓		✓		✓	✓
Face-centred cubic				✓	✓			✓

A simple (primitive) cubic cell gives no systematic absences, whereas the body-centred cubic cell shows only those reflections for which $(h + k + l)$ is even; the face-centred cubic structure gives reflections only from planes where h, k, and l, are either all odd or all even. This information enables you to use a series of values of $\sin^2\theta$ (from a powder photograph) to decide to which type of lattice the numbers belong. For example, if the values of θ are such that the ratios of $\sin^2\theta$ for the first three reflections fit most closely the integers 3:4:8, then the lattice is of *face-centred cubic* type [3 has h,k,l, of 1,1,1), 4 has h,k,l of (2,0,0,), and 8 has h,k,l (2,2,0)]. The ratios are quite different for primitive or body-centred cubic structures.

This procedure can be exemplified by reference to Figs 6.3 and 6.4; you should be able to conclude by inspection at this stage that molybdenum has a body-centred cubic structure, whereas copper is face-centred cubic.

The photographs are reproduced to scale (180 mm = 180°), so that you should also be able to verify that the appropriate θs (measured from the film — the angle between a line and the centre of the film is 2θ) give integral ratios of $\sin^2\theta$ as predicted. You can also calculate the unit cell dimension a.

Ionic crystals

The powder photograph of crystalline sodium chloride (page 78) provides a good example of the effect upon the diffraction pattern when more than one type of ion is present in the lattice. It can be seen from the diffraction pattern that the (1,1,1) reflections are weaker than those for the (2,0,0) planes. This observation can be understood in terms of the intensity of the scattered beam of X-rays and its dependence on the nature of the atom doing the scattering. The X-rays are scattered by the electrons around the nucleus, which will lead to a different intensity (amount) of scattering from Na^+ and Cl^-. Now for the (2,0,0) type reflections, all the planes causing reflection contain Cl^- and Na^+ ions so that all these ions will contribute to reinforcement [three (2,0,0) horizontal layers are clearly shown in Fig. 6.12]. However, for the (1,1,1) plane of sodium ions, there is a layer of chloride ions in between (see Figure 6.18); the X-rays scattered from each layer are exactly out of phase, but do not cancel exactly because their *intensity* is not the same. Thus a weak reflection is observed.

For KCl the two ions have the same number of electrons and cannot be distinguished with X-rays. Although the KCl structure is the same as that for

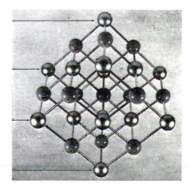

Fig. 6.18 View of the sodium chloride unit cell showing the (1,1,1) planes of one type of ion (indicated) and interstitial planes comprising ions of the other type

NaCl, the (1,1,1) reflection is now actually missing. The X-ray pattern resembles that for a cubic lattice with one half of the spacing of the actual unit cell.

Thus the important information in an X-ray diffraction pattern is contained in the *positions* (values of θ) and *intensities* of the diffracted beams. The former usually allow recognition of the unit cell and lattice type and the dimensions of the unit cell. The intensities of the reflections depend on the *nature* of the ions present and these are the key to the application of X-ray diffraction for structure determination for molecules.

Ionic radii

The measurement of the unit cell length can also lead to values of the ionic radii of the constituent ions. For example, for NaCl the length of the cell side is equal to the sum of the diameters of the sodium and chloride ions (Figure 6.12). In LiCl, which has the same type of structure, the Li^+ ions are so much smaller than the Cl^- ions that the latter actually 'touch' along the diagonal of a face (see Fig. 6.12). From the X-ray powder photograph from this compound, the length of the cell side can be obtained (0.513 nm) and used to calculate the length of the diagonal of the face and hence the ionic radius of Cl^-. Then, from the length of the cell side for NaCl (0.56 nm), you should be able to calculate the ionic radius of Na^+.

From Figure 6.5 you should be able to confirm the face-centred structure of NaCl and confirm the length of the side of the unit cell (0.56 nm).

6.4   The structure of molecules

In a crystal where the lattice points are occupied by covalent molecules (e.g. in a crystalline organic compound) the diffraction from each separate atom must be considered. The molecules themselves will be symmetrically placed with respect to each other, but atoms in the molecules will now not only be found at the corners and body- or face-centres of the cells. However, there will still be characteristic 'repeat distances' in the structure (i.e. periodic variation in electron density) which cause diffraction as discussed before for simple ions, and the essential theory is the same. The X-ray examination is carried out for a single crystal, if one is available, and a very complicated pattern of reflections, with different intensities, is obtained. The problem then is to work back from this information to a plot of the electron density (which causes diffraction) in the unit cell. An electron density contour map, in which the 'peaks' correspond to atoms in the unit cell, can be drawn. For example, Fig. 6.19 shows an electron density map for benzene, and the positions of the carbon atoms (a symmetrical hexagon, with equal C–C bond lengths) can be seen: peaks from hydrogen atoms are not clearly observed because there is relatively little X-ray scattering from these atoms (which depends on the number of electrons round a given atom).

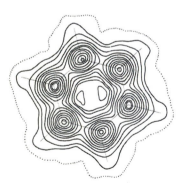

Fig. 6.19 Electron density map for benzene. Reproduced from the paper by E. G. Cox, D. W. J. Cruickshank and J. A. S. Smith, *Proceedings of the Royal Society*, 1958, **247A**, p. 1

Computers make this type of analysis relatively straightforward, and it is now routine to obtain the complete molecular structure of many organic and organometallic molecules, giving a detailed picture of the molecules and accurate measurements of bond lengths and angles. X-ray diffraction also provides an excellent method for investigating the structures of biologically important molecules which contain repeating chemical groups (e.g. proteins and nucleic acids). One important example is the analysis of the X-ray pattern from the nucleic acid DNA, which is interpretable in terms of this molecule having an interwoven double-helix of repeating units of base pairs. X-ray reflections indicate characteristic 'repeat distances' (cf. planes in a crystal) of 0.34, 3.4, and 2.0 nm: these are, respectively, the distance between successive units in the chain, the repeat distance (pitch of the helix), and width of the spiral. This new information enabled the essential features of the DNA structure to be elucidated so elegantly by Watson and Crick in their pioneering and Nobel Prize-winning research.

Computer graphics also enable electron-density maps to be redrawn to give a three-dimensional view of complex structures. For example, Fig. 6.20 shows the X-ray diffraction pattern from a crystal of haemoglobin and a side view of part of the molecule, derived from the electron-density map. The haem ring, containing the large Fe atom with an oxygen molecule bonded above it, can clearly be seen. Those amino acids in the protein chain which are closest to the haem are shown; the histidine residues situated above and

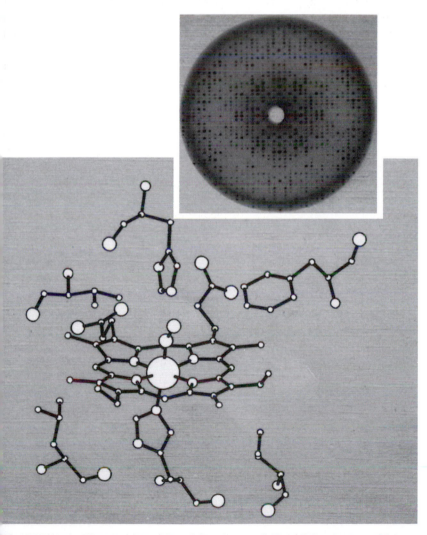

Fig. 6.20 X-ray diffraction pattern (inset) from haemoglobin and the structure of the haem ring, iron atom, and associated amino acids

below the metal can be readily identified. This information is vitally important to the understanding of the biological function of such molecules.

Further reading

W. Clegg, *Crystal Structure Determination*, Oxford Chemistry Primers, 1998.

INDEX

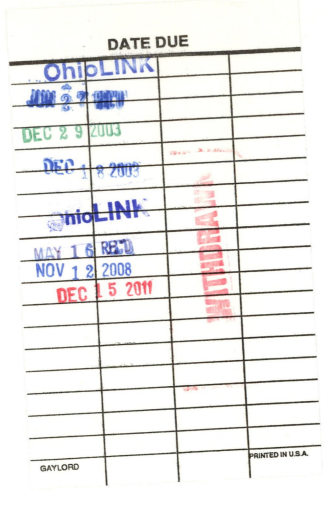